"Only Love Can Save Us From Climate Change"

ONLY LOVE
CAN SAVE US FROM
CLIMATE CHANGE

HOA DAM - Issue 7

Nov. 2019

Published by Lotus Media
Compiled and produced by Hoa Dam Group
Phe Bach & Uyen Nguyen

Content

FOREWORD

It is getting cold and the autumn leaves displace their pretty colors. The new moon is greeting the Earth from the edge of the sky; it is almost in November, and the rain is yet to come. Why hasn't it rained yet? California, both to the south and north, is burning--wildfires are almost everywhere. On the other side of the Pacific Ocean, the Tropical Storm Matmo has just swept through the central provinces of Vietnam. Floods, droughts, and natural disasters are happening everywhere. Climate change is real and it is affecting us all.

According to the United Nations, "Climate change is now affecting every country on every continent. It is disrupting national economies and affecting lives, costing people, communities and countries dearly today and even more tomorrow. Weather patterns are changing, sea levels are rising, weather events are becoming more extreme and greenhouse gas emissions are now at their highest levels in history. Without action, the world's average surface temperature is likely to surpass 3 degrees centigrade this century. The poorest and most vulnerable people are being affected the most."

In this urgent matter, we are asking the same question as Johan R. Platt: "Can Buddhism Help Save the Planet?" The new book, Ecodharma argues that yes, it can — but only if Buddhism saves itself first."

Since one of the noble goals of Buddhism is the alleviation of suffering that ultimately leads to enlightenment, it should be, as

Engaged Buddhism encourages, that it is not only for individual enlightenment, but a broader societal and/or human illumination. The way we move forward positively is lifting each other up.

Thus, the fundamental education in Buddhism is planting-the-seeds and training and practicing: Sowing, nurturing, watering these Bodhi seeds, our tree of happiness and peace will grow. The foundation of Buddha's teaching is the Four Noble Truths and the Eightfold Path. These fundamental truths must pass on to future generations so that each individual as well as the members of the human community live responsibly, peacefully, and in harmony with all people and species as well as in a way that we protect and nurture our Mother Earth.

In the face of climate change, this is a serious issue that needs our immediate attention and action right now. The climate emergency is the defining issue of our times. We, the publishers Hoa Dam and Lotus Media Inc. are compiling English articles, written by many well-known authors, which have been published in international newspapers about climate change. This collection is made for all ages, but especially for our young readers and the Vietnamese Buddhist Youth Families born and raised abroad, in order to raise awareness and take action against the impeccable disasters due to the negative behaviors of humans.

This collection also brings together articles related to the environment and climate change from scholars and practitioners of Buddha's teachings, especially Thay Nhat Hanh, His Holiness the Dalai Lama, and many Lama West. Tibetan, Western scholars, etc.

Let us together practice frugal living, and live in a way that "we offer the joy in the morning, and make life less miserable in the afternoon", live with the Four Ways of Persuasion (Catuh-samograha-vastu) (Tứ Nhiếp Pháp - the Four all-embracing (bodhisattva) virtues: Giving (Dana), Affectionate Speech (Priyavacana), Beneficial Actions (Arthakrtya), Co-operation (Samanartharta), and diligently transform internal and external hurdles and obstacles to become better human beings and leave this beautiful Earth a better place for all.

With that, we would like to introduce this collection, "Only Love Can Save Us from Climate Change".

ASPIRATIONS TO CHANGE THE EARTH?

Dr. Dion Peoples/College of Religious Studies
Mahidol University, Thailand

When I was a child, I spent many years living in the rainy environments of Oregon, Washington and Germany. While living in Germany in the late 70's to early 80's, I became acquainted with acid rain, which was said to have devastated a number of forest-lands in Germany. Then I was relocated to the desert of California, to what is now a closed Air Force Base. Knowing how miserable the California desert is, due to its lack of rain, when I enlisted into the Air Force, I wanted to return to Germany, a place that suits my preferences very well. After leaving the military, I moved to Bangkok, Thailand, to become a Buddhist monk, until I disrobed to pursue higher-levels of education. I've now been in Bangkok for over fifteen years, and have been teaching as a Doctor of Buddhist Studies for over a decade. The point is: I've lived in three different climates for about fifteen years a piece, and can ascertain some differences from those localized perspectives.

I took my older teenage son to the bookstore a few days ago, and I noticed a small book by Greta Thunberg, the young lady who sailed to the UN from Europe, to give a speech about the perils of climate change. Her speech in September of 2019, has already been published and available in the Siam Paragon's branch of Kinokuniya in Bangkok a month later. The main point of her speech, even according to a death-metal/black-metal version of her speech found on social-media sites, emphasized: "How Dare You!" – which I find very telling, and the point of my article here.

"How dare you!" – is directed to the capitalists who decided to develop products for sale, that devastated the environment, our global ecology – poisoning our air, destroying our soil, troubling our waters, damaging our health, deforesting lands, pushing many fauna and flora into extinction.

There is no doubt that the various business enterprises aspire to improve human capabilities through advancing technologies, but at what cost and for how long? When will companies decide to embark into better ecological practices and greener technologies that will benefit humanity and yet protect or revitalize the earth – bettering our surroundings instead of harming our surroundings, or as I teach my two sons: 'Be helpful, not hurtful" – as a motto to live by.

To be honest, there isn't enough time to settle differences. The time for action is now, or rather our worsening conditions should never have been to allow to decay into the present stage where we face worsening air quality, water-quality, food-quality, living-standards, quality of living, working-conditions, and even our modes of transport – as I once read years ago, from a source I am unable to recollect: our distance and time in traveling to work from home and back isn't a measurement of bettering our standard of life. Why must we drive 20-50 or more kilometers a day in traffic-jams just to earn a living? Why are jobs not found near our residences? Why do places of employment not allow potential-employees to be hired nearby, so as to reduce the carbon-footprint of the population. Why must we risk our lives in traffic, just to earn an insufficient wage?

There is a university near my home, but they don't have a Department of Buddhist Studies, or a Faculty of Religion – or something comparable where I can be possibly employed. Instead, the only potential places for my employment are 25 kilometers away, or my former university which was about 61 kilometer away from my home? Why is it not possible to develop my own Buddhist Studies program within a university near my own home, where I don't have to be worried about traveling, and potentially have a greater quality of life? There should be some government agency which can better assist in finding more suitable employment chances or changes for someone in order to lessen our burdens upon the Earth, since one person cannot have all knowledge of which place of employment to work or which city he can be employed at. No one knows where all the job-openings are, and often people are underemployed or take positions where they are over-qualified to take or are moved into fields where they are not perfect candidates. Thus, there seems to be a problem with some people in some places of employment, because they 'just need a job to support their family', rather than finding the best place to be employed. This becomes a structural problem for society.

When people are under-qualified for their positions, this impacts their decision-making processes in their places of employment, and the society at large eventually sufferings due to the ineptitude within the organization.

The concept of "owning" –perhaps a biblical-idea, is the root of the problem. Western-European or American society seems to think that their deity suggests they have dominion over the species on the planet, and it becomes the human-duty to steward the planet – but seeing the state of the planet today, clearly this has been a mistake. When humans proclaim dominance over something, the relationship becomes symbolic of oppression – there isn't a harmonious relationship. Humans, within the sphere of influence from Buddhism have learned to have a codependent relationship with others and the planet. Although there are not really any Buddhist suttas or verses that suggest if someone cuts down a tree they should replace it with seedlings, so that a new tree can eventually be harvested in the future – the ideology of codependency would suggest that a relationship should be maintained, and replacing something borrowed falls within that framework.

The Agganna Sutta from the Digha-Nikaya suggests that the Earth used to be sweet, but owing to the greed of humans, our conditions have ruined, and once fields became partitioned, more problems arose, generating numerous government ordinances which have now confused our lives. The Kalama Sutta famously suggests not to follow certain elements of logic, and if something is unprofitable, blameworthy, censured by the intelligent, or conduces to loss or sorrow, then the endeavor shouldn't be pursued. This is enough to work with for Buddhist critical thinking skills, to ascertain the conditions of our planet, or our present state of affairs.

Capitalism is profitable for the business owners, but their profits are the workers' withheld wages. There hasn't been a fair sharing of profits, enabling the workers to develop themselves as better people, instead they have been left toiling for subsistent wages, similar to conditions of wage-slavery. Whatever has been the cost-of-living in society, wages are often not up to that standard, which forces people to keep returning to the job or face the tragedy of being unemployed and replaced by a worker who will take even less of a salary, so that the owner can make more profit – of course this widens the gap between the "haves and have-nots". Clearly owners are not willing to embark into technologies or production-processes that improve our global state of affairs. Owners are blamable for the harm that they have caused the planet, which was formerly a nearly pristine place to be, before the industrial revolution. Humans before this need to worry basically about reforestation, irrigation and sanitation – however our civilizations have now advanced beyond those primitive stages of civilization into more advanced standards and expectations. Intelligent people have long proclaimed the dangers of capitalism's assault upon the planet, but often they do not own the means of production, so whatever they may say isn't heeded, or is often ignored, even if the media propagates the warning. It's not in the economic interest of the owner to revert into activities that are not within his financial interest – each owner thinks to become the god of their occupational-field. There is too much greed or ego within the class of the capitalists. Many of these enterprises causes loss and sorrow: loss of the forest-environment and sorrow to the humans and species that have lost out due to the interests of the company imposing itself upon a local population. Imperialism was built upon these forces of domination. Europeans didn't just come over to Southeast Asia to make money, often they brought their priests along to criticize the local traditions, implying that their culture was inferior. There wasn't a deep desire to comprehend and respect the local traditions, there was only the motivation to exploit people for any sort of gain or fame.

With the decrease in forested areas and an increase in hazardous air-quality, and harmful chemicals in our waters and air, what is the future to do, in order to survive?

There are only two simple things I can ascertain: keep 'progressing' and destroy the world, or withdraw from capitalism's interests and live more ecological lives. It's ironic that I write this from my computer in an urban environment, as the air-conditioner runs, and I play music on a stereo, while my teenage son is playing online Madden 20 on a Playstation. I often remind him: when I was a youth, there wasn't any video-games, no internet; we played outside all day. I read real books, I never had to turn on a computer to read something – I image all of the hours of electricity I use enriches someone. I suppose that was the goal of urban planners that Jane Jacobs, in her book about the death and life of great American cities, could have warned us about: people who illogically plan cities, without keeping the people in mind, because they are motivated by money.

Soon, there will be no choice but to revert to simpler ways of living, and it may not be a choice, but a necessity. In this sense, Buddhism could be helpful, but how so?

Buddhism teaches us to manage our suffering. Indeed, someone could determine that this is brainwashing, but that doesn't mean that the brainwashing couldn't be a positive factor towards managing our civil affairs. First of all, we have to acknowledge that there is suffering in our world, indeed, it's here under climate-change. When we know that there is suffering, we can turn our attention towards the origin of that suffering. Often the suffering isn't rooted within ourselves, but rather it comes from some external circumstance. We could modify our perception towards a declining world, but the only thing that changes is our opinion of the enduring crisis. The trash is still there, the poor air quality is still there… companies are still exploiting the planet. There can be a cessation to the suffering – would this involve the public-takeover of those facilities causing harm to local and the global ecology? Would it involve the people taking countermeasures, like wearing gas-masks to survive, and always filtering our water, and always buying "food", if that is really what it is (because scientists want to make meat out of feces[1] or force us to drink cockroach milk[2]) – because urban planners prohibit city dwellers to grow their own foods and herbs under their contrived zoning laws, designed to benefit only those with the ability to gain from restricting freedoms in society. They won't let us have our own freely-grown medicines for our troubles, but have no problem to get us addicted to prescription medicines.

There should be a path away from suffering, but will this answer the global climate crisis within our lifetime? There is the possibility to possess the proper views towards the climate crisis, when the mind bends and tends towards knowledge and vision, so when a

---

[1] https://www.livescience.com/14669-poop-meat-safety.html - accessed on 31 October 2019

[2] https://www.sciencealert.com/scientists-think-we-should-start-drinking-cockroach-milk-superfood - accessed on 31 October 2019

person is established into **proper view** (involving knowledge of suffering, its origin, cessation and path – originating from either the voice of another and wise attention; and assisted by virtue, learning, discussion, serenity and insight), then proper views should be pursued, developed and cultivated.[3] This implies our education and proper technological advances that can add towards a better world, rather than something exploitative with a negative gross outcome. Proper view also knows that to be wise, someone should be skilled in the various elements, sense-bases, and skilled with dependent origination, and know what is possible and impossible[4]) – all of this can be done. If we only cared enough and had the proper discipline to put these ideas into action. Transform philosophy into practice – praxis. We should also have the **proper thought** (involving renunciation from our former ways of living wrongly, non-ill-will, and harmlessness towards all beings and the environment) – these two, proper view and thought or intention, comprise the aggregate of wisdom, which we need to develop the planet into something beneficial for all remaining species. The **proper speech** (refraining from lying, slander, harsh speech, frivolous speech – someone should speak only after an analysis) would involve administrators, business-people, and communicators to address policy and disseminate better policy to the population, so that we all can change our consumption behaviors and domestic behaviors, as we transform our economies. **Proper action** (not taking life, no stealing, no sexual misconduct, no intoxicants),

---

[3] The 117th discourse within the Majjhima-Nikaya details more about this.

[4] The 115th discourse within the Majjhima-Nikaya details more about this.

implies that we free our animals to live in open areas and refrain from killing them. I don't have a problem converting to vegetarianism, but I still consume meat only because it is available. If it wasn't, I could easily just consume vegetables as a replacement – the point being made is that companies don't have to provide meat. If it's not there, no problem, buy something else. A **proper livelihood** (involves the giving up of wrong-livelihood), jobs that support the industries that enjoy killing (military, weapons development, poisons, human trafficking, and other harmful occupations). These three comprise the aggregate of virtue. A **proper effort** (involves the exerting the mind to prevent unarisen evil-states and strive to overcome those while producing wholesome states), would involve recycling, stop buying meats, make better consumer choices, and so on. A **proper mindfulness** (there are four foundations to proper mindfulness, contemplating the body as body, feelings as feelings, mind as mind, and mind-objects as mind-objects – internally and externally, arising or vanishing, presence for knowledge and awareness, and not clinging to the four foundations; attention to signs, unwholesome thoughts arise desire, hate and delusion; give attention to other signs connected with the wholesome, subside the evil, abandon unwholesome, steady the mind, become concentrated and quieted[5]) would involve assessing our wrong ways and determining proper ways to move forward as a species, and as a species responsible to correct the mess we have created. A **proper concentration** (this involves the four jhanas; concentration is the

---

[5] The 119th discourse within the Majjhima-Nikaya details more about this.

unification of mind, the four foundations of mindfulness are the basis for concentration, four-strivings are equipment, repetition, development and cultivation is the development of concentration), involves contemplating all we have done, and having places to facilitate the pondering of what we have done to improve our own life and the lives of other species – we need these green-places to take the carbon dioxide and other harmful agents from our air, and provide us with more oxygen. The math is simple, we need more trees, and less people making harmful decisions. Proper effort, proper mindfulness and proper concentration comprise the concentration aggregate. Someone can meditate, pre-meditate, out-meditate, and mis-meditate – circling back to affect the being of a person. A person with such knowledge and vision could be delivered (liberated) through the developed knowledge which should be now, the arising, development and presence of wisdom. When all is said and done, we can leave a greater legacy by improving our own humanity. We have no more excuses. Our children should never have to generate disrespectful thoughts, like: "How dare you?" – because we are supposed to raise the next generation in better ways, but we should leave a better world for them to live in, and inspire them to also improve the world for their future generations. We should never leave our mess for others to clean up. That would be very harmful and disrespectful.

I'm currently trying to change my life-patterns, and I hope everyone reading this will think about changing their own behaviors and soon begin to change their consumption and economic behaviors as well – but this is also a corporate responsibility towards society, it's not just changing the behaviors of the people without excessive amounts of money. Some of that money can be used to empower others to change the world – as long as those with the means to assist are not greedy, hateful or delusional – three unwholesome roots within our human-consciousness's. Please be inspired and find ways to leave a better legacy for our youth and their youth.

Dr. Dion Peoples/*College of Religious Studies Mahidol University, Thailand*

CAN BUDDHISM MEET THE CLIMATE CRISIS?[*]

David Loy/Lion's Roar
September 18, 2019

David Loy makes clear what Buddhism offers in the face of climate change. From the Spring 2019 issue of *Buddhadharma: The Practitioner's Quarterly.*

It is no exaggeration to say that today humanity faces its greatest challenge ever: in addition to burgeoning social crises, a self-inflicted ecological catastrophe threatens civilization as we know it and (according to some scientists) perhaps even our survival as a species. I hesitate to describe this as an apocalypse because that term is now associated with Christian millenarianism, but its original meaning certainly applies: literally an apocalypse is "an uncovering," the disclosure of something hidden—in this case revealing the ominous consequences of what we have been doing to the earth and to ourselves.

Climate issues are receiving the most attention and arguably are the most urgent, but they are nonetheless only part of a larger

---

[*] Adapted from Ecodharma: Buddhist Teachings for the Precipice, published by Wisdom (January 2019)

ecological crisis that will not be resolved even if we successfully convert to renewable sources of energy quickly enough to avoid lethal temperature increases and the other climate disruptions that will cause.

The climate crisis is part of a much larger challenge that includes overfishing, plastic pollution, hypertrophication, topsoil exhaustion, species extinction, freshwater depletion, hormone-disrupting persistent organic pollutants (POPs), nuclear waste, overpopulation, and (add your own "favorite" here...), among numerous other ecological and social problems that could be mentioned. Most if not all of these disorders are connected to a questionable mechanistic worldview that freely exploits the natural world because it attributes no inherent value to nature—or to us, for that matter, since humans too are nothing more than complex machines, according to the predominant materialistic understanding. This larger view implies that we have something more than a technological problem, or an economic problem, or a political problem, or a worldview problem. Modern civilization is self-destructing because it has lost its way. There is another way to characterize that: humanity is experiencing a collective spiritual crisis.

The challenge that confronts us is spiritual because it goes to the very heart of how we understand the world, including our place and role in this world. Is the eco-crisis the earth's way of telling us to "wake up or suffer the consequences"?

If so, we cannot expect that what we seek can be provided by a technological solution, or an economic solution, or a political solution, or a new scientific worldview, either by themselves or in concert with the others. Whatever the way forward may be, it will need to incorporate those contributions, to be sure, but something more is called for.

This is where Buddhism has something important to offer. Yet the ecological crisis is also a crisis for how we understand and practice Buddhism, which today needs to clarify its essential message if it is to fulfill its liberative potential in our modern, secular, endangered world.

Traditional Buddhism focuses on individual dukkha due to one's individual karma and craving. Collective karma and institutional causes of dukkha are more difficult to address, both doctrinally and politically.

Just as climate change is only part of a much larger ecological crisis, so ecodharma is a small part of socially engaged Buddhism, and indifference or resistance to ecodharma is part of a larger problem with socially engaged Buddhism in the US. In the wake of the Great Recession of 2008 the two largest engaged Buddhist organizations, the Buddhist Peace Fellowship and the Zen Peacemakers, almost collapsed due to severely reduced financial support, and since then they have struggled on—often quite effectively, I'm pleased to add—in much reduced circumstances. Noticeably, however, some other Buddhist institutions are thriving financially. In the last few years, for example, Spirit Rock in Northern California successfully fundraised for a multimillion-dollar expansion program. Noticing this difference is by no means a criticism of that accomplishment, yet the contrast in public support is striking. Serious money is available for some high-profile meditation centers, where individuals can go on retreat, but apparently not for organizations that seek to promote the social and ecological implications of Buddhist teachings.

This doesn't mean that socially engaged Buddhism has failed. In some ways it may be a victim of its own success, in that some forms of service—prison work, hospice care, homeless kitchens, and so on—are now widely acknowledged as a part, sometimes even an important part, of the Buddhist path. Note that this is usually individuals helping other individuals. My perception is that over the last generation Buddhists have become much better at pulling drowning people out of the river, but—and here's the problem—we aren't much better at asking why there are so many more people drowning. Prison dharma groups help individual inmates who are sometimes very eager to learn about Buddhism, but do nothing to address the structural problems with our criminal justice system, including racial disparities and overcrowding. In 2014 the number of homeless children in the US attending school set a new record: about 1.36 million, almost double the number in 2006–2007. Why does by far the wealthiest country in human history have so many homeless schoolchildren and by far the world's largest prison population?

Buddhists are better at pulling individual people out of the river because that is what Buddhism traditionally emphasizes. We are taught to let go of our preconceptions in order to experience more immediately what's happening right here and now; when we encounter a homeless person who is suffering, for example, we should respond compassionately. But how do we respond compassionately to a social system that is creating more homeless people? Analyzing institutions and evaluating policies involves conceptualizing in ways that traditional Buddhist practices do not encourage.

A similar disparity applies to the ways that Buddhists have responded to the climate crisis and other ecological issues. My guess is that most people reading this have so far been little impacted personally by global warming, except perhaps for slightly larger air-conditioning bills. We have not personally observed disappearing ice in the Arctic or melting permafrost in the tundra, nor have we become climate refugees because rising sea levels are flooding our homes. For the most part, the consequences are being felt elsewhere, by others less fortunate. Traditional Buddhism focuses on individual dukkha due to one's individual karma and craving. Collective karma and institutional causes of dukkha are more difficult to address, both doctrinally and politically.

I'm reminded of a well-known comment by the Brazilian archbishop Dom Helder Camara: "When I give food to the poor, they call me a saint. When I ask why the poor have no food, they call me a communist." Is there a Buddhist version? Perhaps this: "When Buddhists help homeless people and prison inmates, they are called bodhisattvas. But when Buddhists ask why there are so many more homeless, so many people of color stuck in prison, other Buddhists call them leftists or radicals, saying that such social action has nothing to do with Buddhism."

Perhaps the individual service equivalent that applies to the climate emergency is personal lifestyle changes, such as buying hybrid or electric cars, installing solar panels, vegetarianism, eating locally grown food, and so on. Such "green consumption" is important, of course, yet individual transformation by itself will never be enough.

Imagine Buddhism as an iceberg where all types of social engagement, including ecodharma, form the tip at the top. Beneath them, but still above sea level, is something much bigger and still growing: the mindfulness movement, which has been incredibly successful over the last few years. Within the Buddhist world, however, it has also become increasingly controversial. Here I will not delve into that debate except to note that although mindfulness practices can be very beneficial, they can also discourage critical reflection on the institutional causes of collective suffering, what might be called social dukkha.

Bhikkhu Bodhi has warned about the appropriation of Buddhist teachings, and his words apply even more to the commodification of the mindfulness movement, insofar as that movement has divested itself of the ethical context that Buddhism traditionally provides: "absent a sharp social critique, Buddhist practices could easily be used to justify and stabilize the status quo, becoming a reinforcement of consumer capitalism." In other words, Buddhist mindfulness practices can be employed to normalize our obsession with ever-increasing production and consumption. In both cases the focus on personal transformation can turn our attention away from the importance of social transformation.

The contrast between the extraordinary impact of the mindfulness movement and the much smaller influence of socially engaged Buddhism is striking. Why has the one been so successful, while the other limps along? That discrepancy may be changing somewhat: an increasing number of mindfulness teachers are concerned to incorporate social justice issues, and the election of Donald Trump has motivated many Buddhists to become more engaged.

Nonetheless, the usual focus of Buddhist practice resonates well with the usual appeal of mindfulness, and both of them accord well with the basic individualism of US society—"What's in it for me?" But are there other factors that encourage this disparity between mindfulness and social engagement? Is there something else integral to the Buddhist traditions that can help us understand the apparent indifference of many Buddhists to the ecological crisis?

The Challenge

A few years ago I was reading a fine book by Loyal Rue, titled *Everybody's Story: Wising Up to the Epic of Evolution*, and came across a passage that literally stopped me in my tracks, because it crystallized so well a discomfort with Buddhism (or some types of Buddhism) that had been bothering me. The passage does not refer to Buddhism in particular but to the "Axial Age" religions that originated around the time of the Buddha (the italics are mine):

The influence of Axial traditions will continue to decline as it becomes ever more apparent that their resources are incommensurate with the moral challenges of the global problematique. In particular, to the extent that these traditions have stressed *cosmological dualism* and *individual salvation* we may say they have encouraged an attitude of indifference toward the integrity of natural and social systems.

Although the language is academic, the claim is clear: insofar as Axial Age traditions (which include Buddhism, Vedanta, Daoism, and Abrahamic religions such as Judaism, Christianity, and Islam) emphasize "cosmological dualism and individual salvation," they encourage indifference to social justice issues and the ecological crisis.

What Loyal Rue calls "cosmological dualism" is the belief that, in addition to this world, there is another one, usually understood to be better or somehow higher. This is an important aspect of theistic traditions, although they do not necessarily understand that higher reality in the same way. While all of the Abrahamic traditions distinguish God from the world God has created, classical Judaism is more ambiguous about the possibility of eternal postmortem bliss with God in paradise. For Christianity and Islam, that possibility is at the core of their religious messages, as commonly understood. If we behave ourselves here, we can hope to go to heaven. The issue is whether that approach makes this world a backdrop to the central drama of human salvation. Does that *goal* devalue one's life in this troubled world into a *means*?

Does Buddhism teach cosmological dualism? That depends on how we understand the relationship between samsara (this world of suffering, craving, and delusion) and nirvana (or *nibbana*, the original Pali term for the Buddhist *summum bonum*). Despite many references to nibbana in the Pali Canon, there remains something unclear about the nature of that goal. Most descriptions are vague metaphors (the shelter, the refuge, and so on) or expressed negatively (the end of suffering, craving, delusion). Is nibbana another reality or a different way of experiencing this world? The Theravada tradition emphasizes *parinibbana*, which is the nibbana attained at death by a fully awakened person who is no longer reborn. Since parinibbana is carefully distinguished from nihilism—the belief that physical death is simply the terminal dissolution of body and mind—the implication seems to be that there must be some postmortem experience, which suggests some other world or dimension of reality. This is also supported by the traditional four stages of enlightenment mentioned in the Pali canon: the stream-winner, the once-returner (who will be reborn at most one more time), the nonreturner (who is not yet fully enlightened but will not be reborn physically after death), and the arhat (who has attained nibbana). If the nonreturner continues to practice after death, where does he or she reside while doing so?

If nibbana is a place or a state that transcends this world, it is a version of cosmological dualism. Such a worldview does not necessarily reject social engagement, but it subordinates such engagement into a support for its transcendent goal, as Bhikkhu Bodhi explains:

Despite certain differences, it seems that all forms of classical Buddhism locate the final goal of compassionate action in a transcendent dimension that lies beyond the flux and turmoil of the phenomenal world. For the Mahayana, the transcendent is not absolutely other than phenomenal reality but exists as its inner core. However, just about all classical formulations of the Mahayana, like the Theravada, begin with a devaluation of phenomenal reality in favor of a transcendent state in which spiritual endeavor culminates.

It is for this reason that classical Buddhism confers an essentially *instrumental value* on socially beneficent activity. Such activity can be a contributing cause for the attainment of nibbana or the realization of buddhahood; it can be valued because it helps create better conditions for the moral and meditative life, or because it helps to lead others to the dharma; but ultimate value, the overriding good, is located in the sphere of transcendent realization. Since socially engaged action pertains to a relatively elementary stage of the path, to the practice of giving or the accumulation of merits, it plays a secondary role in the spiritual life. The primary place belongs to the inner discipline of meditation through which the ultimate good is achieved. And this discipline, to be effective, normally requires a high degree of *social disengagement.*

—"Socially Engaged Buddhism and the Trajectory of Buddhist Ethical Consciousness" *Religion & West, issue 9*

Bhikkhu Bodhi distinguishes between the Theravada understanding of transcendence, which sharply distinguishes it from our phenomenal world, and the Mahayana perspective, which understands transcendence to be the "inner core" of phenomenal reality. Nevertheless, in his view both traditions begin by devaluing phenomenal reality. The question is whether "transcending this world" can be understood more metaphorically, as a different way of experiencing (and understanding) this world. Nagarjuna, the most important figure in the Mahayana tradition, famously asserted that there is not even the slightest distinction between samsara and nirvana: the *kotih* (limit or bounds) of nirvana is not different from the *kotih* of samsara. That claim is difficult to reconcile with any goal that prioritizes escape from the physical cycle of repeated birth and death, or transcending phenomenal reality.

In place of a final escape from this world, with no physical rebirth into it, Mahayana traditions such as Chan/Zen emphasize realizing here and now that everything, including us, is *shunya* (Japanese: *ku*), usually translated as "empty."

Shunyata "emptiness" is thus the transcendent "inner core" of phenomenal reality that Bhikkhu Bodhi refers to. That all things are "empty" means, minimally, that they are not substantial or self-existing, being impermanent phenomena that arise and pass away according to conditions. The implications of this insight for how we engage with the world can be understood in different ways. It is sometimes taken in a nihilistic sense: nothing is real, therefore nothing is important. Seeing everything as illusory discourages social or ecological engagement. Why bother?

The important point here is that "clinging to emptiness" can function in the same way as cosmological dualism, both of them devaluing this world and its problems. According to Joanna Macy, this misunderstanding is one of several "spiritual traps that cut the nerve of compassionate action." According to Macy, to see this world as illusion is to dwell in an emptiness that is disengaged from its forms, in which the end of suffering involves nonattachment to the fate of beings rather than nonattachment to one's own ego. But the Buddha did not teach—nor does his life demonstrate—that nonattachment means unconcern about what is happening in the world, to the world. When the *Heart Sutra* famously asserts that "form is not different from emptiness," it immediately adds that "emptiness is not other than form." And forms—including the living beings and ecosystems of this world—suffer.

Many educated Buddhists today aren't sure what to believe about a transcendent "otherworldly" reality, or karma as a law of ethical cause and effect, or physical rebirth after we die. Some wonder whether awakening too is an outdated myth, similar perhaps to the physical resurrection of Jesus after his crucifixion. So it is not surprising that a more secular, this-worldly alternative has become popular, especially in the West: understanding the Buddhist path more psychologically, as a new type of therapy that provides different perspectives on the nature of mental distress and new practices to promote psychological well-being. These include not only reducing greed, ill will, and delusion here and now, but also sorting out our emotional lives and working through personal traumas.

As in psychotherapy, the emphasis of this psychologized Buddhism is on helping us adapt better to the circumstances of our lives. The basic approach is that my main problem is the way my mind works and the solution is to change the way my mind works, so that I can play my various roles (at work, with family, with friends, and so on) better—in short, so that I fit into this world better. A common corollary is that the problems we see in the world are projections of our own dissatisfaction with ourselves. According to this spiritual trap, "the world is already perfect when we view it spiritually," as Joanna Macy puts it.

Notice the pattern. Much of traditional Asian Buddhism, especially Theravada Buddhism and the Pali canon, emphasizes ending physical rebirth into this unsatisfactory world. The goal is to escape samsara, this realm of suffering, craving, and delusion that cannot be reformed. In contrast, much of modern Buddhism, especially Buddhist psychotherapy (and most of the mindfulness movement), emphasizes harmonizing with this world by transforming one's mind, because one's mind is the problem, not the world. Otherworldly Buddhism and this-worldly Buddhism seem like polar opposites, yet in one important way they agree: neither is concerned about addressing the problems of this world, to help transform it into a better place. Whether they reject it or embrace it, both take its shortcomings for granted and in that sense accept it for what it is.

> *When it comes to the ecological crisis, Buddhist teachings do not tell us what to do, but they tell us a lot about how to do it.*

Neither approach encourages ecodharma or other types of social engagement. Instead, both encourage a different way of responding to them, which I sometimes facetiously call the Buddhist "solution" to the eco-crisis. By now we're all familiar with the pattern: we read yet another newspaper or online blog reporting on the latest scientific studies, with disheartening ecological implications. Not only are things getting worse, it's happening more quickly than anyone expected. How do we react? The news tends to make us depressed or anxious—but hey, we're Buddhist practitioners, so we know how to deal with that. We meditate for a while, and our unease about what is happening to the earth goes way... for a while, anyway.

This is not to dismiss the value of meditation, or the relevance of equanimity, or the importance of realizing shunyata. Nevertheless, those by themselves are insufficient as responses to our situation.

When it comes to the ecological crisis, Buddhist teachings do not tell us what to do, but they tell us a lot about how to do it. Of course, we would like more specific advice, but that's unrealistic, given the very different historical and cultural conditions within which Buddhism developed. The collective dukkha caused by an eco-crisis was never addressed because that particular issue never came up.

That does not mean "anything goes" from a Buddhist perspective. Our ends, no matter how noble, do not justify any means, because Buddhism challenges the distinction between them. Its main contributions to our social and ecological engagement are the guidelines for skillful action that the Theravada and Mahayana traditions offer. Although those guidelines have usually been understood in individual terms, the wisdom they embody is readily applicable to the more collective types of engaged practice and social transformation needed today. The five precepts of Theravada Buddhism (and Thich Nhat Hanh's engaged version of them) and the four "spiritual abodes" (brahmaviharas) are most relevant. The Mahayana tradition highlights the bodhisattva path, including the six "perfections" (generosity, discipline, patience, diligence, meditation, and wisdom). Taken together, these guidelines orient us as we undertake the ecosattva path.

Social engagement remains a challenge for many Buddhists, for the traditional teachings have focused on one's own peace of mind. On the other side, those committed to social action often experience fatigue, anger, depression, and burnout. The engaged bodhisattva/ecosattva path provides what each side needs, because it involves a double practice, inner (meditation, for example) and outer (activism). Combining the two enables intense engagement with less frustration. Such activism also helps meditators avoid the trap of becoming captivated by their own mental condition and progress toward enlightenment. Insofar as a sense of separate self is the basic problem, compassionate commitment to the well-being of others, including other species, is an important part of the solution. Engagement with the world's problems is therefore not a distraction from our personal spiritual practice but can become an essential part of it.

The insight and equanimity cultivated by eco-bodhisattvas support what is most distinctive about Buddhist activism: acting without attachment to the results of action, something that is easily misunderstood to imply a casual attitude. Instead, our task is to do the very best we can, not knowing what the consequences will be—in fact, not knowing if our efforts will make any difference whatsoever. We don't know if what we do is important, but we do know that it's important for us to do it. Have we already passed ecological tipping points and civilization as we know it is doomed? We don't know, and that's okay. Of course we hope our efforts will bear fruit, but ultimately they are our openhearted gift to the earth.

It seems to me that, if contemporary Buddhists cannot or do not want to do this, then Buddhism is not what the world needs right now.

David Loy/Lion's Roar
September 18, 2019

A BUDDHIST PERSPECTIVE ON CLIMATE CHANGE

Kara Holsopple /Allegheny Front
Decemmber 18, 2015

The coalition of people across the world who are urging world leaders to do something about climate change is broader than ever. Earlier this year, Pope Francis issued his landmark statement on climate change. And Buddhist leaders have also signed a letter that urged nations at the UN climate summit in Paris to come to an agreement limiting fossil fuel emissions. Recently, Kara Holsopple caught up with Pittsburgh's George Hoguet[*] to talk more about why Buddhists are taking a stand on the climate change issue. He's an ordained lay person in the Buddhist Order of Interbeing and studies with Zen Master Thich Nhat Hanh. Hoguet also took part in last year's People's Climate March in New York and is a founder of the Earth Holder Sangha, which works on climate change issues from a Buddhist perspective.

Allegheny Front: What are some of the tenets of Buddhism that

---

[*] *George Hoguet is an ordained lay person in the Buddhist Order of Interbeing and studies with Zen Master Thich Nhat Hanh. Hoguet is also a founder of the* Earth Holder Sangha, *which works on climate change issues from a Buddhist perspective.*

point towards the environment or caring for the environment?

George Hoguet: You know, there are a lot of different Buddhist traditions and several different schools of Buddhism. But they all kind of have a core understanding. And probably the biggest understanding is that there's no duality. We are not separate from one another, we are not separate selves. There's more "non-George" in George—you know, there's my parents in me, there's the sun, there's the earth, there's the breath of animals. And that's pretty consistent in Buddhism. The protection of the environment—the Buddha spoke to that himself, and a Bodhisattva is kind of a Buddhist saint. And there's one called The Earth Holder to protect the animals—a lot like St. Francis of Assisi.

AF: Earlier this year, the Pope's encyclical, or teaching on the environment, made a big splash and sets up acting on climate change as a moral issue. Buddhist leaders recently released a letter to world leaders in advance of the Paris climate talks, urging them to stop relying on fossil fuels. That was a big deal, right?

GH: That was a major deal. Right after the Pope issues that encyclical on June 18, within that week, the Dalai Lama endorsed it. I think the Buddhist leaders recognized that this is coming from human activity, and our teacher Thich Nhat Hanh got engaged with this back in the '70s. In 1970, there was a major scientific conference, and they were looking at population growth. We were at 3.5 billion; now we're almost eight. But they also saw fossil fuels coming on. So they speak to that. The Buddhist leaders speak a lot more to the spiritual pollution that we have—the extravagance that we have and not really understanding how our decisions affect other people.

AF: Are you talking about over-consumption?

GH: Certainly Buddhist practices call for a simple lifestyle. There's a call to get as close to vegan as you can because of the impacts of emissions from livestock and dairy. I don't want to say that they want to target any one given industry. They're really asking us to look at where it's all coming from, and it's coming from us not paying attention—not being connected to Mother Earth.

AF: What are Buddhist leaders asking their followers to do about climate change?

GH: Buddha means "The Awakened One," so they're asking us to really wake up and to help wake other people up. Again, my teacher Thich Nhat Hanh is credited with the term "engaged Buddhism." And that came from the fact that he was from Vietnam, and bombs were dropping in the neighborhoods and the monks could not just sit in the monastery while communities were being destroyed. They went out and got engaged. And he feels the same way towards this crisis and this issue. So [that means] encouraging other people without pointing blame, without making "the Other" other, because then the conversation can't go.

AF: Do you think science and politics need religion to get the support they need for action on climate change?

GH: I think where we need the moral aspect is—where the religions will speak to—is this idea of sharing and not considering others "Other;" and recognizing that in order for the technology to move forward, that we will need to be able to be in relationship with one another.

AF: And how does your practice of Buddhism inform your environmental work? Or, maybe the opposite, how does your environmental work inform your faith?

GH: When I look at what just happened in Paris, and what took us so long to get there, it seems to me that we are finally speaking to one another. I was a lot more caustic about things we needed to do, and who are the bad guys, and all of that, for a while. And I found that just puts people off. It doesn't really help bring people together. The solutions are going to be found by everybody coming together, so my teachings and my practice, which is about equanimity, really inform that a lot. I think my environmentalism is very consistent with the teaching of Buddhism and certainly our teacher, who's written many articles about Mother Earth. So I think they work hand-in-hand. But it's kind of calmed me down a little bit and helped me listen to the other side.

Kara Holsopple /Allegheny Front
Decemmber 18, 2015

CAN BUDDHISM HELP FIGHT CLIMATE CHANGE?

Lucia Graves/Pacific Standard
Oct 2, 2018

At an idyllic retreat in California, the architect of the Paris Agreement argues that it can.

Amid the golden hills near Point Reyes, California, in the sunlit main hall of the Spirit Rock Meditation Center, Christiana Figueres, the architect of the Paris climate agreement, is explaining how Buddhism saved her life.

Her talk is part of a daylong gathering of activists, yoga instructors, Buddhist practitioners, and meditation enthusiasts all intent on bringing more mindfulness and loving kindness to their approach to climate activism. Timed to coincide with the Global Climate Action Summit in San Francisco hosted by Governor Jerry Brown, Saturday's retreat is about an hour's drive from the city—and a world away.

While the dress code at the summit in the city was "business," there are no shoes allowed here. And when I ask if I can take my purse inside instead of leaving it in an open cubby by the entrance, a custodian smiles sympathetically and says, "You *can*," before launching into a "funny story" about that time his expensive sunglasses went unmolested in a cubby here for four whole days.

I smile back at him. I take my purse.

The day's featured speakers at this famed meditation retreat include climate diplomats like Figueres, the former head of United Nations climate negotiations in Paris, but also Tibetan-Buddhist scholars and activists like Julia Butterfly Hill, the woman who lived in a redwood tree for 738 days to keep it from being cut down.

Figueres is in the middle of explaining how, a few years ago, when she was working on the pathway to Paris, she experienced the most difficult personal trauma of her life. "I thought, 'I wonder what would happen if I just disappeared at this point,'" she tells the seated crowd of shoe-less climate activists.

Instead of giving up, she reached out to a friend.

"I said: 'I'm suicidal. I have this responsibility. I can't do this. I have to do something,'" Figueres recalls. "He says, 'What do you want to do?' And I say, 'Buddhism.' And he goes: 'Buddhism? What do you know about Buddhism?' And I say: 'Nothing. In fact, I'm not sure I even know how to spell it correctly.'"

Her friend then turned Figueres onto the teachings of Thích Nhất Hạnh, a Vietnamese Zen Buddhist monk whose books have become popular in the West. "The teachings of Thích Nhất Hạnh saved my life," Figueres says, but, more importantly, "they were the guiding light" for her work on the Paris Agreement, helping her muster the strength, compassion, and focus she needed to do the job.

It wasn't just a personal salve, Figueres insists. Thinking like a

Buddha, she says, could help ordinary citizens put their climate ideals into action too. "Finding strength in our pain at the individual level is what we need to do at the global level," she says.

Americans are in what one speaker calls "a moment of awakening consciousness." Specifically, with respect to climate, it's a moment of recognizing that the task of protecting the planet can't be left up to politicians. (After Donald Trump announced his intention to withdraw the United States from the Paris climate agreement, it became clear that the current administration can't be counted on for much.)

Jack Kornfield, co-founder of Spirit Rock, sees this moment of awakening as in line with what Thích Nhất Hạnh has said about how "the next Buddha" might not take the form of an individual, but rather of "a community practicing understanding and loving kindness."

"What's beautiful is the empowerment of people," Kornfield says. "It's also problematic," he adds wryly, "because it means you. That's the downside. Otherwise you can offload the responsibility to the spiritual leaders or the climate leaders."

The uses of hope; the uses of fear

Figueres speaks of finding strength in her pain, but what does that actually mean?

The emotions typically associated with climate change are fear and anxiety. On the topic of climate messaging, some critics have questioned the wisdom of dwelling on the negative, and have called for more hope in how we talk about our impending doom.

And earlier that week at the Global Climate Action Summit,

many of the speakers had seemed intent on splitting the difference.

"It almost puts us in a place of schizophrenia," Johan Rockström, executive director of the Stockholm Resilience Centre, says of the current state of the climate crisis. "There's never been a reason to be so nervous as today based on the scientific necessity, but there's never been so much reason to be hopeful."

Los Angeles Mayor Eric Garcetti coined a term for this contradictory emotion: *anxcited.* "It's like every day, this feeling of *anxcitement,*" Garcetti told a gaggle of reporters at the summit when asked about his mood. "Wow, we are finally there and we can do this. And it's getting worse simultaneously."

Political change often starts with anger, which is an animating force in activism and something Figueres called for explicitly at a kickoff event to the GCAS, saying, "We have to get to the point of public outrage."

This Saturday at Spirit Rock, though, no one appears motivated chiefly by animus, or by negative emotions more generally. One speaker even recalls being actively disgusted to hear a political operative say that their "approach is to infuriate and disgust our base so they will go out there and vote." It feels like a cheap trick, the speaker says—manipulating negative emotions (though of course it's also true that voters can be moved by a righteous and deeply justified anger).

"It's important to be able to feel our pain and grief," says Spirit Rock co-founder James Baraz. "And it's also important to, when they occur, maintain and increase those wholesome states."

The approach doesn't mean denying emotions like anger as they occur. It means not being derailed or directed by them exclusively.

And today it means trying to look at climate change as though the glass is half full.

Paraphrasing the writer Gary Snyder, Kornfield says: "If you're going to save [the world], don't save it out of fear and anxiety. Save it because you love it."

Mindfulness goes West

Meditation as a practice has been around for thousands of years, but it has seen a striking rise in modern America, from its introduction through beatniks and hippies to the New Age movement and, more recently, the mainstreaming of yoga and meditation. The surge in interest is not just happening in California—although Californians would have you know that it was happening here first.

"California is a very powerful place because we are the trendsetters for the whole world," Tibetan Buddhist scholar Anam Thubten tells the assembled. "We have to be really careful what we do because everyone's going to follow us."

In fact, everyone's already doing so. Mindfulness techniques are now used in schools and the military. Fortune 500 companies offer them to improve employee well-being and productivity. And Ohio Congressman Tim Ryan, who's weighing a bid for president and wrote a book about meditation, is saying he wants to cultivate "the yoga vote." A majority of Americans now view Buddhism favorably, according to Pew Research Center data.

Though often subsumed into a multibillion-dollar wellness industry, most mindful practices, including yoga and meditation, were at least originally intended to be free. In the West, of course, even breathing can be monetized. The popular meditation app Head Space turned "peace of mind" into a $250 million business last year.

Attendees of Spirit Rock's "Loving the Earth: Healing the Planet Through Mindful Engagement" today have paid between $60 and $200—a sliding scale at the payer's discretion, with an additional $10 penalty for anyone failing to carpool. All proceeds are donated to the green groups that helped organize the proceedings.

The day's spiritual instructions include the advice, "Don't *should* all over yourself," a quote from Julia Butterfly Hill, who shares other lessons she learned firsthand from her time in the tree.

"We look at these challenging times and think, *This is too much*," she says, of the daunting nature of the climate crisis. "But every action is changing our world, moment by moment." Awakening to that means not asking whether you can contribute, she insists, but how.

"I happen to have been well-designed to be the woman who lived in a tree for two years and eight days," Hill says. "No matter what our unique gifts are, there is a way to access them and make a difference."

Later, when we're prompted to make a personal climate commitment and say it aloud, David DeSante, the white-haired ornithologist sitting next to me, offers a refreshingly non-technical idea: "To link arms with others."

DeSante is the founder of the Institute for Bird Populations in

Marin County, and will happily chat for an hour about bird brains or what he loves so much about thrushes—did you know they can produce two independent sounds at the same time, harmonizing with themselves? When it comes to climate change though, he's learned not to beat people over the head with "the club" of science, and today he says he's less interested in offering answers than in finding allies.

The mind-body connection

The day's activities focus not just on the mind, or even on the heart (which is Spirit Rock's preferred organ for thinking). They also incorporate the body at every turn.

A session on "mindful movement" takes place in a nearby meadow. During lunch, we're invited to explore the retreat center's many pathways into the hills. Before eating, we're reminded to do so slowly, savoring every bite.

The tip isn't an exercise in hedonism—not strictly. A review of research from an eight-week training program developed by Jon Kabat-Zinn at the University of Massachusetts Medical School found that Kabat-Zinn's Mindfulness-Based Stress Reduction Program, famous for its mindful meditation on raisins, "is an effective treatment for reducing stress and anxiety" associated with daily routine and chronic illness.

The imprimatur of science often trails what's anecdotally apparent to the discerning observer—willow bark tea was popular during childbirth long before aspirin was an option—but, unlike that $55 Rose Quartz egg claiming to "intensify" your femininity, scientific studies confirming meditation's benefits keep rolling in.

Under the right conditions, mindfulness can reduce pain, help manage stress, ward off depression, and slow down or even reverse neurodegeneration.

Can it help solve climate change though? That null hypothesis has yet to be disproven.

But Figueres, for her part, thinks it is a crucial tool—and not just as a balm for the people who, like her, run high-level climate talks.

"The Paris Agreement is not about me," she says. "It's about the emergence of this true love and solidarity and recognition that we're all here together."

Lucia Graves/Pacific Standard

Oct 2, 2018

WHAT WE CAN LEARN FROM BUDDHIST INSIGHTS FOR THE CLIMATE CRISIS

Sister True Dedication, Contributor/
Buddhist nun in Zen Master Thich Nhat Hanh's
International Plum Village Community
11/04/2015 05:50 pm ET - Updated Nov 04, 2016

On October 29 our Teacher, Thich Nhat Hanh, came together with His Holiness the Dalai Lama and Buddhist leaders from over a dozen countries to make a common declaration on climate change to world governments ahead of the Paris Climate Summit this December. It is a continuation of his tireless engagement to bring compassion and insight to the climate crisis, and builds on his powerful message "Falling in love with the Earth," submitted to the United Nations in 2014.

"We are on course to self-destruction," said Thich Nhat Hanh, speaking in France in June last year. "Many civilisations on Earth have been destroyed, because they could not live in harmony with nature. If we continue like this there is *no doubt* that our civilisation will be destroyed. Earth may take many millions of years to heal, to re-establish her balance and restore her beauty. For the Earth, it's not a problem. But for us humans it's a *big* problem. We will disappear from Earth."

"We need more than just new technology to protect the planet," he says. "We need real community and co-operation. We need to re-establish true communication — true communion — with ourselves, with the Earth, and with one another as children of the same mother."

The international declaration also includes the signatures of over 40 Plum Village Dharma Teachers, monastic and lay, from the U.S., Europe and Asia. The Plum Village International Community of Engaged Buddhists, founded by Thich Nhat Hanh over 40 years ago, is the largest community of its kind in the world, with over 1,300 local groups and over 800 monks and nuns. Every year at Plum Village practice centres and retreats, tens of thousands of people are trained in "the art of mindful living" in harmony with one another and the Earth. Over 100,000 people have made a commitment to follow Thich Nhat Hanh's Five Mindfulness Trainings, a simple, practical ethical code for daily living in the 21st Century that has the power to heal and transform not only our individuals live and family life, but also our society and the planetary situation.

"Everyone can do something," says Sister Chan Khong, the most senior nun in the Plum Village community. "We can make a difference, starting today. We can eat 50 percent less meat. That is a big act of compassion for ourselves, for the planet, and for the little cow, sheep, pig or chicken. If we look deeply and contemplate the beauty and dignity of these animals when they are alive, the suffering they endure in the abbatoir, and the destructive impact of industrial meat farms on the Earth, compassion will be born in our heart. Reducing the amount of meat we eat is not an obligation or duty. It arises naturally out of love."

"What we do as an individual is deeply interconnected with the collective. Each one of us can make a promise to do one thing to help. We can turn off extra lights, not use plastic bags, or eat vegetarian a few days a week. We are part of the bigger change. A little love can become Great Compassion, for all people and for all beings. The act may be small, but it contains our intention of love and mindfulness of what is happening to our precious home." It is in this spirit that our international community will be coming together to support the Global Climate March on November 29th, and to send compassionate energy for the success of the Paris Climate Summit.

For decades, Thich Nhat Hanh has been a leading international figure calling for a radical shift in the way of thinking and acting to address the situation of climate change. He helped convene the ground-breaking Dai Dong Environment Conference of Scientists back in 1970 (a key stimulus to the UN's first ecology summit), triggered a global initiative of national "No-Car Days" under the auspices of UNESCO in 2006, and has published bestselling books on Buddhism and ecology, The World We Have(2008) and Love Letter to the Earth (2013). In 2007 he led tens of thousands of his followers in shifting to a vegan diet. "In the past Buddhists were vegetarian with the intention to nourish our compassion towards animals. Now we know we should be vegan in order to protect the Earth."

In recent years Thich Nhat Hanh has contributed a powerful

new spiritual dimension to the climate crisis, one that has inspired UN Climate Chief Christiana Figueres. "We need to change our way of thinking and seeing things," he has said. "We need to wake up and fall in love with Earth." The deepest root of our current crisis is, he says, "the wrong view that you and the Earth are two separate entities: you think the Earth is only the environment. You are in the centre and you want to do something for the Earth in order for you to survive." But this, he says, "is a dualistic way of seeing things. It does not have the insight of interbeing. The Earth is so much more than simply your environment." This insight of interbeing is a unique Buddhist contribution to the climate change crisis. Love, compassion, generosity, and the insight of our mutual interdependence are, he suggests, what is needed — both in December's talks and long into the future. "Only when we truly love the Earth will our actions spring from reverence and the insight of our interconnectedness. That's the kind of awareness, the kind of awakening that we need. The future of the planet depends on whether we're able to cultivate this insight or not. It is our love for the Earth that will give us enough compassion, strength and insight to change our way of life."

Living simply, and staying connected to ourselves, our loved ones, and the Earth is key. "Centuries of individualism and competition have brought about tremendous destruction and alienation," he says. "Many of us are lost, isolated and lonely. We work too hard, our lives are too busy, and we are restless and distracted, losing ourselves in consumption. We buy and consume things we don't need, putting a heavy strain on both our bodies and the planet." Thay describes what he calls "the spiritual pollution of our human environment": the toxic and destructive atmosphere we're creating with our way of consuming, in the films we watch, the news or magazines we read. These, he says, all-too-often water the seeds of fear, anger, greed and hatred in our consciousness. He encourages us to consume in such a way that keeps our compassion, peace and generosity alive. "We don't need to consume *a lot* to be happy; in fact we can live very simply. With mindfulness, any moment can become a happy moment. Savoring one simple breath; taking a moment to stop and contemplate the bright blue sky; or fully enjoying the presence of a loved one, can be more than enough to make us happy."

During the height of the Vietnam War, Thay reminded us that "Man is not our enemy. The real enemy is our ignorance, discrimination, fear, and violence." The enemy now is not our governments or corporations, but the energy of greed, mindlessness and indifference. If in the Vietnam War we needed to stop shooting each other, in our current planetary crisis we need to stop our rampant culture of consuming and exploiting. It is a challenging time because the front line is not to be found in the battlefield, but within ourselves. We must take responsibility for our part, in our way of living and in our choices in each moment. Only then can we say that we love the Earth. Only then will we have a chance.

Sister True Dedication, *Contributor/Buddhist nun in Zen Master Thich Nhat Hanh's International Plum Village Community*
11/04/2015 05:50 pm ET - Updated Nov 04, 2016

AWAKENING IN THE AGE OF CLIMATE CHANGE

David Loy/Tricycle
Spring 2015

If Buddhism is to address the ecological crisis, it must clarify its essential message.

Let me begin by emphasizing what most of us already know about climate change. First, it's the greatest threat to human civilization ever, as far as we can tell. Second, it's not an external threat but something we are doing to ourselves. And third, our collective response remains, if not completely negligible, very far from adequate.

Yet climate breakdown is only part of a much larger eco-crisis. We cannot blame the degradation of nature simply on recent increases of carbon in the atmosphere. If we are to avert climate disaster and our own potential extinction, we must address our long-standing degradation of the natural world in all its forms. Humanity has been exploiting the natural world for most of its existence. Today, however, business as usual has become a threat to our very survival.

E. O. Wilson, the renowned Harvard biologist, predicts that by the end of this century about half of all the earth's plant and animal species will become extinct or so weakened that they will disappear soon thereafter. Scientists tell us that there have been at least five other extinction events in the earth's history, but this is the fastest ever and the only one caused by the activity of one particular species: us.

The whole eco-crisis attests to the fact that we are a globalizing civilization that has lost its way. The crisis of nature is, at heart, a crisis of civilization. Shifting to renewable sources of natural energy will not by itself resolve our collective preoccupation with never-ending economic growth—and the often meaningless production and consumerism it entails—that is incompatible with the finite ecosystems of the earth. Many things could be said from a Buddhist perspective about why this fixation on growth cannot provide the satisfaction we seek from it, but let's take a look at one particularly revealing example: what Mitsubishi is doing with bluefin tuna.

The Japanese love sashimi, and their favorite variety is bluefin tuna. Unfortunately, bluefin tuna is also one of the world's most endangered fish. But the Mitsubishi conglomerate, one of the world's largest corporate empires, has come up with an ingenious response: It has cornered close to half the world market by buying up as many bluefin tuna as it can as the worldwide population plummets toward extinction. The tuna are imported and frozen at -60°C in Mitsubishi's massive freezers, for they will command astronomical prices if, as forecast, Atlantic bluefin tuna soon become commercially extinct as tuna fleets try to satisfy an insatiable demand—primarily Mitsubishi's.

From an ecological standpoint, this response is immoral,

obscene. From a narrow economic standpoint, however, it's quite logical, even clever, because the fewer bluefin tuna in the ocean, the more valuable Mitsubishi's frozen stock becomes. And it's the nature of economic competition that corporations like Mitsubishi are sometimes encouraged or "forced" to do things like that: if you don't do it, someone else probably will. That's how the "tragedy of the commons" plays out on a global scale.

The example above is one of many that point to a fundamental perversity built into economic systems motivated by profit, which tend to devalue the natural world into a means, subordinated to the goal of expanding the economy in order to maximize profits. This focus often overshadows our appreciation of the natural world, which means that we end up destroying real wealth—a flourishing biosphere with healthy forests and topsoil, oceans full of marine life, and so on—in order to increase numbers in the bank accounts. As the enormous gap between rich and poor continues to widen worldwide, most of that increase goes into a very small number of accounts.

Such perverse logic ensures that sooner or later our collective focus on endless growth—on ever-increasing production and consumption, which requires ever more exploitation of our natural resources—must inevitably run up against the limits of the planet, and it just so happens that's happening now. Today it's not enough for us to meditate and pursue our own personal awakening; we also need to contemplate what this situation means, and how to respond.

Many Buddhist teachings are relevant here, especially their emphasis on interdependence and nonduality. We consider ourselves and others to be separate entities, pursuing our own well-being at the cost of theirs in ways that the eco-crisis repudiates. As earth-dwellers, we're all in this together. When China burns coal, that pollution doesn't just stay above Chinese skies, nor does nuclear radioactivity from Fukushima stay only in Japanese coastal waters. The same is true generally for humankind and the rest of the natural world: when the ecosystems of the earth become sick, we become sick. In short, the ecological crisis is also a spiritual crisis: we are challenged to realize our interdependence—our larger "self"—or else. What the earth seems to be telling us is *Wake up or get out of the way.*

From this perspective, the problems that challenge us today are even more intimidating. Facing seemingly intractable political and economic systems, we could easily despair. Where to start? Those who control our current economy and political systems also profit the most from them (in the narrow sense), so they tend to be little inclined to make the systemic changes necessary—and are often incapable of doing so.

We can see that institutional change can only come from the grassroots, and signs are growing that more and more people are fed up with waiting for economic and political elites to take action. As the author and environmentalist Paul Hawken points out in his 2007 book *Blessed Unrest*, a vast number of large and small organizations are working for peace, social justice, and sustainability—perhaps two million, he now estimates. This is something that's never happened before: it's as if the organizations have "sprung up" from the earth to act as its immune system, responding to the cancer that now threatens our survival.

But while the necessary response has begun, it's easy to overlook what's happening, because the mainstream media are not interested in publicizing or encouraging that transformation. Six megacorporations now control 90 percent of the media in the United States, and they make their profits not from informing us but from advertising. Their perspective inevitably tends to normalize consumerism, including the political system that aids and abets it. Unsurprisingly, they promote "green consumerism" as the solution to the eco-crisis—personal lifestyle changes such as buying hybrid or electric cars, installing solar panels, eating locally, and so on. As Bill McKibben has pointed out, however, even if many of us do everything we can to reduce our individual carbon footprints, "the trajectory of our climate horror stays about the same." But if the same number of us work all-out to change the system, he continues, "that's enough."

The ecological crisis, and the larger civilizational predicament of which it is a symptom, is just as much a crisis for the Buddhist tradition, which needs to clarify its essential message in order to fulfill its liberative potential in the modern world.

Related: No Easy Answers

One of the important developments in contemporary Buddhism has been socially engaged Buddhism, and *service*—prison dharma, hospice work, helping the homeless, and the like—is now widely accepted as an important part of the Buddhist path. Buddhists have become much better at pulling drowning people out of the river, but—and here's the problem—we're not any better at asking why there are so many more drowning people, or what's pushing them into the river upstream.

I'm reminded of Dom Hélder Câmara's famous quote: "When I

give food to the poor, they call me a saint. When I ask why the poor have no food, they call me a communist." Is there a Buddhist version? When Buddhists help homeless people and prison inmates, they are called bodhisattvas; but when Buddhists ask why there are so many more homeless, so many rotting in prison, other Buddhists call them leftists or radicals. "That has nothing to do with Buddhism," the others say.

At the same time as Buddhist organizing for social and economic justice has floundered, the mindfulness movement has seen incredible success. Mindfulness offers an individualistic practice that can fit nicely into a consumer corporate culture focused on efficiency and productivity. Although such practices can be very beneficial, they can also discourage critical reflection on the institutional causes of collective suffering, or *social dukkha*. As Bhikkhu Bodhi has warned: "Absent a sharp social critique, Buddhist practices could easily be used to justify and stabilize the status quo, becoming a reinforcement of consumer capitalism."

Recently I read a passage in *Everybody's Story: Wising Up to the Epic of Evolution*, by Loyal Rue, that stopped me in my tracks, because it crystallizes so well a discomfort with Buddhism (or some types of Buddhism) that has been bothering me for some time. Rue writes that religions such as Christianity and Buddhism will keep declining as it becomes increasingly clear that they can't address the great challenges facing us today. He cites two basic problems: *cosmological dualism* and *individual salvation*, both of which "have encouraged an attitude of indifference toward the integrity of natural and social systems."

Cosmological dualism is obviously an important aspect of

Christianity, one that distinguishes God in his heaven from the world he has created. But Buddhism also dualizes insofar as this world of samsara is distinguished from nirvana. In both traditions, the contrast between the two worlds inevitably involves some devaluation of the lower one: so we are told that this realm of samsara is a place of suffering, craving, and delusion. And in both cases, the ultimate goal is *individual salvation*, which involves transcending this lower world by doing what is necessary to qualify for the higher one, whether that is eternity in heaven with God or attaining nirvana.

One can point to aspects of the Buddhist tradition that do not support cosmological dualism, especially the famous statement by Nagarjuna, the influential Buddhist philosopher and founder of the Madhyamaka school, that "there is not the slightest difference between nirvana and samsara." Yet that claim must be balanced against (for example) the early Buddhist doctrine that nirvana involves the end of physical rebirth, or the Mahayana Pure Land schools that contrast this world with Amitabha's Pure Land.

Buddhists don't aim at heaven: we want to awaken. But for us, too, salvation is individual: yes, I hope you will become enlightened also, but ultimately my highest well-being—my enlightenment—is distinct from yours. Or so we have been taught.

When it comes to the nature of enlightenment, however, most of us aren't sure what to believe. Since many modern Western Buddhists reject the idea of rebirth, it is not surprising that a *this-worldly* alternative has become popular in the West, where understanding the Buddhist path as a program of psychological development helps us cope with personal problems, especially our "monkey mind" and afflictive emotions. This has led to innovative types of psychotherapy as well as the recent success of the mindfulness movement, which represents the culmination of this trend in Western Buddhism. Buddhism is providing new perspectives on the nature of psychological well-being and new practices that help to promote it—reducing greed, ill will, and delusion here and now, for example, but also sorting out our emotional lives (not a big issue in Asian Buddhism) and working through personal traumas.

Related: The Buddha's Footprint

This development has been largely beneficial, but it has a shadow. The common presupposition of the more secular Buddhism is that my basic problem is the way my mind works, and the solution is to change the way my mind works, so that I can play my various roles (work, family, friends, etc.) better, so that I *fit into this world better.* Most of Asian Buddhism is concerned with escaping this world, since *samsara* can't be changed, but for much of contemporary Western Buddhism, the path is all about changing myself, because *I'm* the problem, not the world.

So while traditional Asian Buddhism emphasizes *ending rebirth* into this unsatisfactory world, much of Western Buddhism, including most of the mindfulness movement, emphasizes *harmonizing* with this world. That means neither is much concerned about social engagement that works to change our world; both take the world (including its ecological crisis and social injustice) for granted, and in that sense accept it as it is.

Both approaches encourage a different way of reacting to the eco-crisis: ignoring it. When we read or think about what is happening, how do we react? We become anxious, of course, but Buddhists know how to deal with anxiety: we meditate, and our unease about what is happening to the earth goes away—for a while, anyway. Needless to say, that is not an adequate response.

The point here is that Buddhist difficulty with social and ecological engagement can be traced back, in part, to this ambiguity about the nature of awakening. And this ambivalence is a challenge we can't keep evading: we really do need to clarify what the essential message of Buddhism is.

There is an alternative way of understanding the Buddhist path, one that is not reducible to the either/or of escaping this world or simply harmonizing with it. The path of personal transformation is about deconstructing and reconstructing the self, or, more precisely, the relationship between the self and its world. Because my sense of self is an impermanent psychosocial construct, with no reality of its own, it is always insecure, haunted by *dukkha* [suffering] as long as I feel separate from the world I inhabit. We usually experience this as a sense of lack: something is wrong with me, something is missing, "I'm not good enough." Consumerism encourages us to perceive the problem as a personal lack: I don't have enough money, I'm not famous enough, attractive enough, and so on. Buddhist practice helps us wake up from this bad dream.

A really important social implication of this deconstruction and reconstruction of the self brings us back to social engagement, including eco-dharma, the application of Buddhist teachings to our ecological situation. As we start to wake up and realize that we are not separate from each other, nor from this wondrous earth, we also begin to realize that the ways we live together, and the ways we relate to the earth, need to be reconstructed as well. That means not only social engagement as service, but finding ways to address the problematic economic and political structures—the institutionalized forms of greed, ill will, and delusion—that are deeply implicated in the eco-crisis. Within such a notion of liberation, the path of personal transformation and the path of social transformation are not really separate from each other. We must reclaim the concept of awakening from an exclusively individualistic therapeutic model and focus on how individual liberation also requires social transformation. Engagement in the world is how our personal awakening blossoms.

It just so happens that the Buddhist tradition provides a wonderful archetype that can help us to do that: the bodhisattva. We overcome deep-rooted self-centered habits by working compassionately for the healing of our societies and the healing of the earth. This is what's required for the Buddhist path to become truly liberative in the modern world. If we Buddhists can't do that, or don't want to do it, then Buddhism might not be what our world needs right now.

David Loy/Tricycle
Spring 2015

AWAKENING FROM CLIMATE SLUMBER

Linda Heuman/Tricycle
Spring 2019

Can Buddhist theology help save us from climate disaster? The Dalai Lama thinks so.

Humanity faces an environmental crisis so critical that our survival on earth is in peril. Yet we have another even more urgent problem: most of us go on living as though nothing out of the ordinary is happening. What is wrong with us? Is there something religions could do to spur us to action? If so, what? Is there a role Buddhists could play? To address these questions, in 2011, at the request of the His Holiness the Dalai Lama, Mind & Life Institute convened a think tank of more than a dozen leading scientists, interdisciplinary scholars, and theologians at his private residence in Dharamsala, India. Other Buddhist luminaries joined in, including Roshi Joan Halifax, Thupten Jinpa, Matthieu Ricard, and His Holiness the 17th Karmapa. This meeting was the twenty-third Mind & Life dialogue with the Dalai Lama. As is typical of these dialogues, it provided a forum both to educate the Dalai Lama and to solicit his input; it also provided a rare venue to introduce Buddhist perspectives to cutting-edge interdisciplinary scholarship, interfaith dialogue, and public discourse. (Atypically, it focused on a topic unrelated to cognitive science.) *Ecology, Ethics, and Interdependence: the Dalai Lama in Conversation with Leading Thinkers on Climate Change*—a skillfully edited and easily readable transcript of the dialogue—shares that important conversation with a wider audience. (Interested readers can also watch videos of the meeting on YouTube.)

Related: Reflections on an Impermanent World

"We're all in a kind of trance," says meeting moderator and book co-editor Daniel Goleman in the opening chapter. "Today, we're facing a real paradox: even though we love our children as much as anybody in human history has, every day, each of us unwittingly acts in ways that create a future for this planet and for our own children, and their children, that will be much worse." The struggle to understand, articulate, and address this paradox is a central theme of the book. "I feel that Buddhism and Christian theology, as well as philosophy and psychology, have very important perspectives to offer science," Goleman continues. "Science documents what's happening, but it doesn't necessarily have within it the mechanisms to mobilize people to act in a skillful way."

The book comprises ten lectures and the ensuing discussions. It is divided into three parts: the first part lays out the scientific evidence showing the consequences of human activity on the planet; the second weighs the ethical implications of environmental change; and the third explores effective action. This weeklong conversation traverses a wide range of questions from the factual (What is happening, to whom, and how fast?) to the philosophical (Do future humans have rights?) to the practical (Should we be vegetarians?). It introduces creative metaphors to facilitate action—such as "handprint," a counter-balance to environmental "footprint," which offers a way to quantify the positive impact of our actions. Or "mindprint," which adds intention to the calculation. And it strikes a hopeful tone, not only focusing on what humans are doing wrong but also providing examples of where we are doing it right— such as in Bhutan, where 50 percent of the country is national park, farming is headed toward all-organic, and carbon emissions are on decline.

Environmental scientist Diana Liverman starts off the presentations, summarizing the latest scientific findings on the state of the environment. By means of a series of charts, she demonstrates the growth in human activity and resource use since 1950 and the corresponding increase in environmental degradation—a phenomenon known as *the Great Acceleration*. The acceleration results from two factors: the number of people on the planet (good news: expected to level off at nine billion by 2050) and how much each person consumes (well, that's bad news). Scientists are concerned about multiple environmental tipping points, she explains; if we cross these thresholds, change will be rapid and irreversible. In this context, climate change is merely one of nine looming catastrophes including chemical pollution, ocean acidification, and biodiversity loss.

Related: How a Growing Buddhist Movement Is Responding to the Ecological Crisis

We have moved from the Holocene into the *Anthropocene*, Liverman explains, a new geological epoch in which the primary shaping force of the planet is human activity. The emerging understanding of the earth is that all its multiple systems—land, oceans, atmosphere, and living things—and all its multiple processes—physical, chemical, and biological—are entangled in complex ways with each other and with human life and activity. "It is very important to remember that the earth's system is not separate from us, but rather that we are part of it," she emphasizes, introducing a second central theme of the book: interdependence.

Climate science evolves quickly. Between the occurrence of the meeting and the publishing of the book seven years have elapsed, so potential readers might fear the content is dated. Alas, recent facts are no less concerning, so the fundamental issues addressed by the book are still relevant. More complicating for the reader is that the political backdrop of the climate conversation has inverted in the United States since the meeting occurred; now climate-change-deniers govern one of the world's leading polluters. In this sense, the book sometimes reads as a relic of a qualitatively different and more hopeful age. For example, after Liverman finishes her talk, the Dalai Lama offers advice that seems naive in the face of today's political reality. "Global leaders should be exposed to this kind of data so that people who are responsible for countries will become fully convinced of the seriousness of the situation. More awareness needs to be created, particularly awareness in free countries where leaders are chosen through election."

In his repeated insistence that education is the solution to public inaction, the Dalai Lama acts as a foil. His is the voice of common sense. Get people the information about what is happening; explain that changing their way of life is in their own best interest; and of course they will comply. Except that isn't what happens, the Dalai Lama's interlocutors are quick to inform him, explaining that the findings have become politicized, forces of active *dis*information are at work, and even when receptive people get the memo, often they don't respond. (The fact that the commonsense approach fails is an important indication that something is wrong with contemporary common sense, an insight that emerges chapters later.)

If peoples' failure to respond is not caused by a lack of data, then what *is* its cause? We have a design flaw in our brains, Goleman suggests, offering a perspective from evolutionary psychology. After all, we have brains "designed for detecting snarling tigers, not the very subtle causes of planetary degradation," he said, so the danger simply doesn't trip the neurological alarm system. The panelists consider many other possibilities as the dialogue advances. Science is telling people things they don't want to hear; the implications are too severe. Or: the future is too far away; we aren't emotionally moved by it. Or: people aren't convinced by data; statistics just aren't the sort of thing that convinces.

Throughout these exchanges, the Dalai Lama demonstrates a faith in science education laudable for a religious leader but misguided in its scientism. "In these modern times, the scientist can sometimes be considered a guru, a person of authority on these issues," the Dalai Lama insisted. "The gurus need to come out and speak." Or later: "I think the best people to stimulate awareness about what's happening and what needs to be done are not the politicians or leaders but the scientists. They are the real gurus in these matters." When the issue at hand is environmental change, certainly natural scientists are the best authorities. But when the matter concerns motivating public response, they are not. The Dalai Lama here is making three common—but telling—category mistakes. First, he conflates all knowledge with knowledge of natural science. Motivation is not a matter of natural fact; it is a matter of human meaning and values, which is the forte of experts like theologians, qualitative scientists, and humanists. Second and third, he misappraises the kind of awareness at stake and the corresponding kind of response needed—thus missing game-changing distinctions that enter the conversation only when Christian theologian Sallie McFague presents. To introduce these distinctions, McFague asks her audience to imagine giving up driving or flying. "The shock that we feel when we imagine this causes us to realize how far we have to go in our attitudes and our practices. We human beings are so embedded in the culture of consumerism that being asked to consume less makes us almost gasp. And we do; we stop for a moment, and then we have to inhale and take another breath, and get back in our cars and our airplanes, and continue on."

That feeling of mind-stopping shock is an important indication that peoples' lack of response is not operating at the level of explicit *knowledge* (which thereafter is referred to as "awareness"); it is operating at the level of implicit *assumptions* (thereafter "deep awareness"). McFague puts it like this: "The culture of consumerism is not just a form of life that we can accept or reject. It has now become like the air we breathe, and this is the nature of culture. Culture becomes nature; it becomes natural. It becomes the way things are; it becomes the world in which we live."

Anthropocene is ostensibly a scientific term referring to an epoch in *natural* history, but it has another important meaning in the context of the humanities, which went unmentioned in Dharamsala. It refers to the corresponding era in *human* history. This meaning of "Anthropocene" is harder to get your head around: it is the era during which nature has been, all along, something humans were assuming it was not. The notion captures an uncanniness and shock, suggests writer Amitav Ghosh, similar to the moment "when a harmlessly drifting log turns out to be a crocodile." (See Ghosh's book *The Great Derangement: Climate Change and the Unthinkable*, which is an excellent companion read.)

Buddhism contains parallels. For example, when we learn the truth of suffering, we might imagine we have acquired knowledge about the world "out there"—that it is in the nature of suffering—but what really matters is what we learn about *ourselves*: we had imagined otherwise; we had it wrong. And gaining that insight is not a matter of acquiring a new fact. Rather, it is relinquishing an illusion.

Throughout the industrial revolution and beyond, modern people assumed that nature was separate from humans, inert, predictable, without agency, under our control, and there for our exploitation. That's how we ended up in this mess. But the escalating environmental crisis acts as incontrovertible counterevidence to these assumptions. We now know that nature has never been like that. We had it wrong.

Understanding that there is a connection between nature and humans reveals a further surprise. We had it wrong on not just one but two counts; that is, about both nature and humans. In light of science's new conception of nature, continuing our out-of-control use of resources is illuminated as logically incoherent, morally wrong, and existentially absurd. But questioning consumer ideology—"the culture of insatiable greed"—in turn undermines our very identity. We no longer know who we are in the most fundamental scheme of things, where we fit into the big picture as human beings, or what we should be doing, McFague observed. "Change at this level is very, very difficult, and in fact, most people find it impossible."

We cannot solve the crisis with the paradigm that created it. And this is why education is necessary but not sufficient to awaken public response. Succeeding at that lies not in playing this game better, as it were, but in playing an entirely different game. We need a paradigm shift.

To identify the kind of shift needed, we have to examine the nature of consumer culture. In what manner are we entrenched? What kind of change would abandoning it entail? McFague says, "Consumerism is a cultural pattern that leads people to find meaning and fulfillment through the consumption of goods and services. Given this," she posits, "consumerism is the newest, the latest, and the most successful *religion* [italics mine]."

To adequately address the planetary crisis then, it would seem we need a culture-wide transformation akin to a *spiritual awakening*. Indeed, McFague uses such language when she says, "We need to wake up to the lie held in the current worldview of individualist, selfish fulfillment... We need to wake up to a different worldview, one that shares all our resources with our fellow creatures." Such an "awakening" sounds comparable in scope to a *religious conversion.* "The change has to happen at all levels of our life," she confirms, "personal, what we eat, how we get to work, taxes, car emissions, everything."

In responding to McFague, the Dalai Lama pinpoints what is at stake. There are theistic and nontheistic religions, he observes, "but we need a third religion." As the chapters progress, conference participants sketch an outline of what that "religion" might look like. (Dalai Lama: "One without scriptures, that is based simply on common sense, our common experience, our inner experience, warmheartedness, a sense of concern for others' well-being, and respect for the rights of others.") Thupten Jinpa, the Dalai Lama's primary interpreter, contributes a Buddhist mechanism for facilitating a value-system shift. And the Karmapa adds his personal environmental conversion story, conveying what environmental awakening might look like from a first-person perspective. Creating a new social consciousness to bring about an alternative future might seem like a brazen undertaking, but cutting-edge theorizing of this very sort is already well underway in academic fields like science studies and the environmental humanities. Moving forward, it could be fruitful to connect the dialogue that emerged from Dharamsala more explicitly with that broader conversation.

A reader might begin *Ecology, Ethics, and Interdependence* imagining that the role for religions in motivating public response is a peripheral one—on the order of planting trees in monasteries, finding sutras that support environmental ethics, or preaching conservation from the pulpit. But by the end of the book, it is clear that is not the case. In the vision that comes forth from the conference, religion occupies center stage.

And in this project, Buddhism could play a leading part. After all, don't Buddhists have experience developing a culture based on a kind of waking up? And how did the Buddha do it? He articulated the problem; identified its causes; established that the problem could be fixed by abandoning its causes; and taught a step-by-step path for doing that. Then he created a community and evolved institutions to support people undergoing that transformation. Might this example not serve as a parallel for how to awaken humanity from climate slumber? The Buddha provided a strategy to recognize and escape from a predicament so existentially dire that we are encouraged to respond as though our hair is on fire. Today, as the Dalai Lama says, in the sentence that closes the book, "The earth is our home, and our home is on fire."

Linda Heuman/Tricycle
Spring 2019

A BUDDHIST'S REACTION TO CLIMATE CHANGE

Ven. Thubten Jampa/Abbey Outreach, News
Nov 30, 2016

While the US was experiencing a difficult and divisive presidential campaign, marked by turmoil and confusion, major changes were happening, and are still happening, in the Arctic, Tibet, the jungles of Borneo, the Gulf of Mexico and other wildlife habitats around the world. In November 2016, the *Washington Post* reported that the North Pole was experiencing temperatures 36 degrees Fahrenheit warmer than normal. Warmer temperatures slow the formation of the ice pack, cause melting of ice in the Polar Regions, leading to a rise in sea levels and a decline in wildlife such as polar bears.

Recent Reports

Dr. Silvan Leinss from the ETH College in Zurich (a leading international university for technology and natural sciences) announced in September 2016 that two major glaciers collapsed in Western Tibet. Tibet is sometimes called the Third Pole because it stores the most ice and water after the Arctic and Antarctic. The effect of melting glaciers in the Himalayan region will have an impact on the supply of drinking water for millions of people and will have a major impact on the hundreds of millions of people living on low-lying land in Bangladesh.

Also just recently in autumn 2016, the World Wildlife Fund published an article indicating the animal population has declined 58% since 1970, and if nothing changes the population will continue to decline up to 67% by 2020. As a result, animals like the lynx are, or will be, threatened mostly due to human encroachment, industrialization, and environmental pollution, and in the case of other species such as leopards, whales and rhinoceros due to their use in clothing, food, and medical production.

There are many more examples of major changes in our environment related to disturbance of the elements (water, fire, earth and wind) due to climate change. It's obvious that major changes are happening in our environment and governments. Citizens worldwide are becoming more aware and are increasingly more determined to slow the advance of global warming. There are many positive steps in that direction.

Positive Steps

Around the same time as the US elections this year, the COP (Conference of the Parties) organized the 22nd Climate Change Conference in Marrakesh. This conference has been proclaimed as the peak of practice, the more practical approach after the Paris COP 21 in 2015. The Marrakesh conference was organized to make sure the commitments made in 2016 become reality.

Our friend Gary Peyton, an environmental advocate from Idaho, shared with us the outcome of the Paris Climate Change Conference in 2015. He told us 195 countries adopted a global agreement to combat climate change and to intensify the actions and investments needed for a sustainable low-carbon future. Furthermore, this conference brought a deep understanding of the intensity of human impact on the planet.

Now in November 2016, 15,000 delegates met again in Marrakesh. They worked hard to translate the promises they made in Paris into work schedules and deadlines. The election in America, of course, has been watched closely, but so far the delegates have not been greatly influenced by it. The trend to a climate friendly environment is already on its way as is the determination to make sure that global warming will not extend 1.5 to 2 degrees Celsius. In the U.S., many counties, cities, and communities have already made steps toward an energy revolution, exiting from dependence on nuclear and fossil-fuel energy.

For example, the U.S. has the most active nuclear plants (100), but it's reducing this number. The plans are to turn off 30 nuclear plants during the years of 2019/20 and only four new plants are under construction in Georgia and South Carolina. The U.S. uses quite a good amount of alternative energy, 11.7%, and there are increases in the use of solar and wind energy.

But the most impact this year in Marrakesh has been made by strong commitments from 47 developing nations. Most countries in Africa, joined by Costa Rica, Sri Lanka, and Bangladesh, agreed to increase their use of alternative energy sources such as sun, wind, bio energy, and water power up to 100% by 2050!

So there is a lot of activity on the governmental level, but that very much depends on the economics and on people—on each of us here in this room, in New York, Texas, Florida, Germany or Singapore and worldwide.

Resources for Buddhists

I have written a couple of articles that you can find on thubtenchodron.org on the topic of environmental protection and general engagement. I have shared how a healthy relationship with our environment on different levels can be established. I really recommend that you look it up, if you are interested: "The Earth is Our Only Home."

Also I have translated a meditation to raise consciousness for a healthy relationship with nature that my teacher Geshe Thubten Ngawang wrote more than 20 years ago: "Meditation to Raise Consciousness for a Healthy Relationship with Nature." This meditation helps to develop more compassion and a sense of our interdependence with our environment, with each and every sentient being on this planet as well as plants and the four elements—water, fire, earth, and wind.

The Abbey's Environmental Efforts

I would like to especially point out what the Abbey has been doing for several years to protect the environment to increase our own awareness and to hopefully inspire you to follow in our footsteps:

First of all, I want to emphasize the Buddha was a founder of environmentalism. His *Vinaya*- the code of behavior that he established for monks and nuns—emphasizes that, "*You cannot cut trees, you cannot cut leaves; you cannot cut flowers; you cannot disturb the forest; you cannot foul the river; you cannot foul the grass.*" In this sense the Buddha instructed his Sangha to observe the same guidelines environmentalists are striving for.

Although we monastics at the Abbey take care of the forest by cutting trees and digging in the ground, we do it in order to improve the environment and to support a healthy forest with a good amount of biodiversity. Our hopes are that the Abbey forest serves as a green belt for wildlife and a firebreak for the area. With the reality of climate change, we are paying extra attention so that the trees and habitat will be healthy and strong enough to survive.

Also, we remove noxious weeds to encourage the return of natural grasses to our meadows. We provide safe habitats for animals by creating a "no hunting" zone in Abbey lands, and we have replaced barbed wire throughout the property with wildlife-friendly fencing.

In addition, we do many daily things to care for our land and the environment such as:

- Saving water by taking short showers.
- Recycling plastic, glass, metal, paper, and so on.
- Driving into town only when necessary or running many errands at once.
- Using geo-thermal technology to heat and cool our buildings.
- Reducing electricity through simple acts like turning off lights when not in the room or turning off computers when not in use.
- And, of course, we do not eat meat.

There is an interdependent relationship between our afflictions, negative emotions, and how we relate to our environment in order to gain wealth and satisfy our senses, we pollute the environment, the earth, water and air.

Working Together

We can do everything in our power to heal the earth, starting with ourselves, reducing our own afflictions and being satisfied with what we have. Also, aspirational prayers and meditation helps to stir our awareness in this direction. Prayers are not just blind faith! They give our mind very positive imprints and direct our energy and actions to specific goals.

Also we need to learn about ecology and train ourselves to come to an understanding how to stop global warming. Many environmental advocates or organizations can provide excellent information on ecology and environmental engagement.

I would like to end with an aspirational prayer by Thrangu Rinpoche, a renowned Kagyu teacher and the personal tutor of HH the 17th Karmapa.

An Aspiration Concerning Global Warming
by Thrangu Rinpoche

May the blessings of the exalted sources of refuge,
The Buddha, his teachings and community: the Three Precious
Jewels,
And the spiritual teacher, meditational deities and protectors of the
Buddhist teachings: the Three Roots,
Fully pacify the terrors of illness, famine and war,
Along with chaotic disturbances of the four elements—
The imminent and terrifying danger that the whole world will
become a great wasteland,
As temperature imbalance causes the solid glaciers of snow mountain

massifs to melt and contract,
Afflicting rivers and lakes, so that primeval forests and beautiful trees
near their deaths!

May the sublime endowments of good fortune and spiritual and
temporal
well-being flourish,
And may all beings nurture one another lovingly and kindly,
So that their joy may fully blossom!
May all their aims be fulfilled, in accordance with the sacred
teachings!

Ven. Thubten Jampa/Abbey Outreach, News
Nov 30, 2016

5 BUDDHIST PRACTICES TO HELP TACKLE CLIMATE CHANGE

Lama Willa B. Miller/Lion's Roar
September 23, 2019

Climate change can feel so immense that it hurts just to think about. Lama Willa Miller offers five meditations to help bring the truth of climate change into your awareness and lay the ground for a skillful response.

Hurricanes and wildfires have come and gone, leaving hundreds dead. We're left facing a dire reality: we live on a warming planet. Homes blown apart. Lives lost. Ecosystems flattened. This is how climate change arrives at our doorstep.

With the destruction comes a wider acceptance of the scientific reality — and a growing motivation to contribute to solutions. But destruction also brings despair, fear about the future, grief, and panic. As we grapple with our new reality, contemplative practice can offer techniques for holding these challenging truths.

Spiritual practices are not alternatives to swift, wise action. They are complementary disciplines to education and activism. Spiritual resources can help us move from desperation to sustainable activism.

How do we get from anger to compassion?

Spiritual practice may not provide concrete climate solutions, but they do have the potential to shift consciousness. Practices and teachings can address how we relate to our grief, despair, and fear. These resources help restructure our understanding of what it means to be human, now, on our home planet.

Here are five tried and true contemplative practices from the Buddhist tradition that can help us hold the truths of climate change, species extinction, and the ecological crisis in our hearts and minds. While this list of practices is not by any means exhaustive, it is a beginning. Even though their roots are ancient, these practices are timely as we encounter the truth of suffering on a global scale.

1. Find a grounding in ethics

Some people see climate change as an ecological issue. Some see it as an economic issue. Some see it as a social issue. But, we know that human actions are at fault. In this sense, climate change is an ethical issue.

Our beliefs about justice — the values that we hold most dear — form the bedrock of our actions. These values are largely learned and assimilated from our culture. Each of us — as individuals and communities — can influence the values upheld by our culture.

Climate change is happening because of what we have valued and how we have conceived of our identity as human beings on this planet. The values have come from a dominant industrial ethos. Climate change, therefore, isn't just a matter of what we *can* do. It's a matter of what we *should* do.

Contemplative traditions teach moral reflections on our actions, speech, and thought. The Buddha emphasized ethics, *śila*, as a fundamental training for his monks. His monastic code of ethics was constructed around the idea of *ahimsa*, or non-violence. Essentially, the Buddha taught that ethical actions are those arising from a commitment to non-harm, gentleness, and simplicity.

Buddhism and other religious traditions have long identified love and compassion as motivators that drive effective and sustainable action.

If we extend śila to our relationship to land, water, natural resources, and animals, non-harm, gentleness, and simplicity become points of reflection for change-making.

Later Buddhist traditions developed rules of conduct, oriented towards compassion, such as the Bodhisattva precepts. These precepts extend from the idea that *bodhicitta,* or wise compassion, is the ground of ethical action and speech. We too can ground our activism, social engagement, and resistance in wise compassion. We can make our activism not about what we are working against, but what we are working for. If we place our activism and relationship to the earth squarely among our deepest values and beliefs, we are more likely to turn again and again to the issue — not out of obligation, but out of genuine commitment.

2. Get comfortable with uncertainty

If there is one thing that climate scientists agree on, it is that we don't know for certain what will happen as the earth warms.

Evidence indicates that tipping points and crises cannot be averted. We have no how idea how much we can slow or ameliorate the suffering. We do not even know how long our species — and others — can survive changes that destabilize the conditions necessary for life. We are stepping into the void.

We want to know if our children and grandchildren will be able to visit the shoreline, walk in the forest, breathe clean air, and live in safety. It is human to fear that the world as we know it may be ending. This uncertainty can feel deeply unsettling.

Many of the Buddha's teachings focus on uncertainty, not as an inconvenience, but as a source of liberation. The Buddha taught that nothing is certain, because nothing transcends impermanence. He called impermanence a "mark of existence" — an undeniable truth of what it means to be alive. To encourage his monks and nuns to face their mortality, he sent them to meditate in charnel grounds — open-air cemeteries — where they could witness decaying corpses.

The Buddha was not trying to torture his disciples. He was trying to free them. While awareness of our mortality stirs our deepest fears, it also frees us from the chains of attachment that bind us. The loosening of attachment helps us open to the truth that nothing is certain. Nothing can be taken for granted. This is how we learn to love the truth for what it actually is.

There is good reason to embrace the uncertainty of climate change as a liberating practice. The more we fear uncertainty, the more likely we are to avoid thinking about climate change. In fact, our worst enemy might not be climate denial, but rather a subtle, subconscious rejection of climate change, based on our fear of the unknown.

If, however, we embrace the truth of uncertainty, we can develop the courage to stay open and engage with the world. If we can accept the fragility of life on earth, we can invest ourselves in the possibility of collective action.

3. Work with emotions

Along with the discomfort of uncertainty, climate change can evoke many other difficult emotions. Witnessing ecosystem destruction and mass extinction, we respond with grief and sorrow. Encountering denial and global apathy, we experience anger. When we consider our children's future, we experience trepidation and worry.

> *Anger can be a protective energy, a healthy response to that which threatens what we love.*

Recently, I was talking to a European graduate student who was writing her thesis on the power of stories to affect climate change. The primary motivator for her work, she told me, has been anger.

Understandably, fear and anger often fuel activism. These primal emotions have kept us alive for centuries. They are good short-term motivators when we are in immediate danger. However, fear and anger are poor long-term motivators. Eventually, they result in stress and burnout — the insidious undoings of activists.

So, we need other chronic motivators for our work. In this area, spiritual traditions have much to offer. Buddhism and other religious traditions have long identified love and compassion, for example, as motivators that drive effective and sustainable action. The *bodhisattva*, a Buddhist archetype of compassion, typifies the possibility that positive and constructive emotions can be the primary fuel for activity. But how do we get from anger to compassion?

Tibetan Buddhism teaches that the states that we most wish to avoid are actually the key to our freedom. Instead of erasing emotions, we can metabolize them. If we take our reactivity into a contemplative space, it is possible to liberate the energy of emotion, transforming it into supple responsiveness.

We might start with an emotion like anger. When anger is heavily fixated on an object, it becomes isolating, contracted, and draining. When we take anger into a contemplative space, we can lighten our focus on the object and the story, turning inward to consider the emotion itself and our part in it.

When we take responsibility for our own anger, we can find its upside. Anger is not always reprehensible. It can be a protective energy, a healthy response to that which threatens what we love. That insight itself can liberate reactive, contracted anger into its deeper nature, a wiser, more inclusive resolve to act with decisiveness and courage in the interest of love.

In contemplative practice, anger can become an inspiration for empathy. We discover that uncomfortable states, while they belong to us, are not to our's alone. Many others also feel anger, including the people we have othered. When we recognize that *this is how so many others feel,* we can commune with the suffering of others. We redirect our attention from the story stimulating anger to our empathy for all those impacted by climate change — even the deniers. By redirecting our focus from a polarizing narrative to a uniting one, we start building a more sustainable platform for action.

4. Access new wisdom

In discussions about climate change, we seem to primarily access one way of knowing — the intellect. The climate issue is couched in the language of conceptual knowing. This conceptual approach — typified by Al Gore's documentary, *An Inconvenient Truth* — is critically important. We need to know what is happening, and why.

However, our response will be much more powerful and resilient if we begin to access other ways of knowing, transforming conceptually-motivated activism into an activism of the heart.

There are two alternative ways of knowing that Buddhist practice and meditation generally rely on: bodily wisdom and non-conceptual wisdom.

BODILY WISDOM

To encounter our human body is to encounter the natural world. We tend to forget that we are mammalian primates! The closer we come to the body, the closer we draw to the truth of our own wildness. This connects us to the planetary wildness that we aspire to protect.

While the mind is tugged into the past and future, the body is fully present. The body's present wakefulness is one of its great wisdoms, and we can easily access that wisdom. It is as close to us as this moment's inhale and exhale. While we want to stay mindful of creating a sustainable future, we don't want to do that at the expense of missing our life. The body reminds us that we are here, now, and our presence is our most powerful resource.

NON-CONCEPTUAL WISDOM

Buddhist meditation also introduces us to the life beyond the conceptual mind — non-conceptual ways of knowing. The wider truth is that human experience is not just mental content. While we spend a great deal of time enmeshed in our world of ideas, there is more to the mental-emotional life than what we think and believe. There is a non-conceptual space in which all of this content arises, and that space can be sensed and widened through the experiences of body. In the practice of the Great Perfection, this space is identified as naked awareness, a part of our mind that is just experiencing, prior to forming ideas about our experiences. The space of awareness can be cultivated until it becomes a holding-environment for relative issues such as climate change.

We can make our activism not about what we are working against, but what we are working for.

As we begin to identify with non-conceptual space, we access a non-dual mode of perception. In the non-dual mode of perception, the illusion of separateness is perforated. This illusion of separateness may be one of the root causes of the crisis we are in. When we are caught up in that illusion, it becomes somehow okay that my consumption happens at your expense. If we are to live sustainably, we need to get used to the idea — nay, the reality — that we are all intimately connected. Meditation leads us there.

5. Find community

A friend of mine once attended a City Council meeting in her local community and ran into a woman who was repeatedly raising the issue of banning plastic bags. Discouraged, the woman said that she could not seem to earn the respect of the city council. My friend replied: "You don't need respect. You need a friend. One person is a nut. Two people are a wake-up call. Three people are a movement."

That friend was the environmentalist and author Kathleen Dean Moore, and her story inspired me. A small, committed group of people can change the world, as Margaret Mead said. Finding a community of activists might not be as daunting as we might think. It can be as simple as finding a few like-minded people and starting a conversation.

In order to gracefully lean into the challenges that we face as a planet, community is critical. But it also does double-duty, laying the foundation for spiritual life.

The Buddha's close attendant Ananda once inquired of his teacher, "Surely the *sangha* [spiritual community] is half of the holy life?"

The Buddha answered, "No, Ananda, do not say such a thing. The sangha is not half of the holy life. It is the whole of the holy life."

The Buddha felt very strongly about the power of community to support the path to awakening. He lived most of his life in intentional community, and identified sangha as one of the three spiritual refuges, along with the teacher and the dharma.

Now is a good time for the eco-curious in the dharma world. There is a growing community of people who seek both spiritual development and activism. If you are one of those people, now especially, you need not despair. Your people are out there.

As we are propelled forward by the consequences out of a warming planet, it is more important than ever that activists and contemplatives work together. We can benefit from an exchange of technologies. While I have highlighted five spiritual technologies to help contemplate climate change, activists have other tools and perspectives that can assist spiritual communities to take action. Activist communities have resources for education and technologies of peaceful resistance that can help contemplatives enact change.

While we grapple with the effects of climate change, we will need tools of resilience and inner work. As dharma practitioners, we bring essential gifts to the project of healing our world. Our challenge is to bring these gifts to bear and continue their development.

By practicing with ethics, uncertainty, emotion, wisdom, and community, we develop an intimate understanding that being human is about what we think and what we believe — and we deepen our ability to embody our work.

Embodiment sends an indelible message that peace and sustainability can become a lived reality. Even when they are imperfectly realized, we can inspire the sense that our lives have meaning, and that we are living our way into ever-increasing integrity with — and service to — our beautiful, unfathomable and sacred world.

Lama Willa B. Miller/Lion's Roar

September 23, 2019

ENGAGED BUDDHISM
AND THE ENVIRONMENT
A Solution to Slow Down Global Warming

Dr. Phe Bach/Mira Loma High, SJUSD,
Sacramento, CA.
Dr. Khanh T. Tran/AMI Environmental, USA.

According to the majority of researchers, the state of global warming or climate change is very real and is gradually and severely growing in both the size and the danger they pose. The main reason of this is the excessive increase in the emission of carbon dioxide within the past 30 years due to the burning of fossil fuels (Cox, P.M., et al., 2000), as well as other chemical components that are not organic (Hansen, J., et al., 2000), as well as due to the living necessities of human beings; from the release of toxic wastes, smoke, and polluted air in industrial facilities of all scales, to automobiles, deforestation, livestock farms, waste materials coming from hydroelectric dams, coal plants, nuclear power wastes, etc.

Lorenzoni, I., & Pidgeon, N. F. (2006), have pointed out that, "The effects of humanity, if left unchecked, on the climate system could create dangerous changes harmful to other aspects that are directly interconnected to the survival of other species on this entire Earth".

Venerable Thich Nguyen Hiep in his 'The Ethics of Buddhism and the Environmental Dilemma', also wrote:

"The human world always has to face natural-born disasters such as floods, earthquakes, volcanic eruptions, tsunami waves... these are the problems they often have to deal with throughout the span of their history of development. And today, the severity of those events are gradually becoming worse due to many negative factors created by humans. Aside from the usually seen natural disasters, the pollution of the air, the decline of underground water sources, spoiled lands, desertification, the change of climate and the disappearances of natural biomes, are the horrifying catastrophes that we, the human race, are facing as well. Those disasters are happening everywhere, everyone saw them, but the needs of living standards and economic development are causing humans to mistreat them, neglecting all the catastrophic risks that they will one day be exposed to."

Take Vietnam as an example, this Southeast nation also has to take responsibility in this decaying state of global climate change. According to the World Population Review (2015), the population of Vietnam had reached 94.5 million people, taking the 14th highest position in the world, accounting for 1.33% of the human population on Earth. Vietnamese people are becoming more and more numerous, but our resources are deteriorating day by day. There are no longer "golden forests, silver seas" like in the old time when we went to the village school. The pollution dilemma is serious, stemming from the technological development, the abuses of authority, the lack of systematic consciousness, and the absence of environmental protection policies, etc. At this moment, in Vietnam, the gaps between the rich and the poor are growing, and the extravagant lifestyle of the rich (the 1%) has been worsening the entire environment in Vietnam.

Therefore, we must be mindful of the problems regarding the environment, even in the smallest of things, such as public hygiene, to larger complications, like manufacturing technologies that contribute to climate change, because what we do will affect many generations to come.

Venerable Tam Phap in 'Buddhism and the Environment', the Buddha Enlightenment selection, various authors, mentioned in Part III: "The Buddha came to be and became enlightened for nothing other than the sympathy towards the common folks who suffered from the ravaging Three Poisons of ignorance, aversion, and attachment. Because of the greed to fulfill material needs, humans must suffer from catastrophic pain. To end the suffering, humans must live in accordance to the path of enlightenment, which is to live by the laws of nature or the law of causes and effects. By this law, humans, floras, and faunas co-exist in a mutual and correlative relationship. Nature shall provide the environment for humans and animals to live in. In return, human beings must have the consciousness to protect nature to maintain a healthy and balanced ecological environment."

More than ever, every single one of us must bear the responsibility and duty for our children's and grandchildren's generations, especially when we were born Vietnamese. Our love and concern for our country are the responsibilities and duties of every child of Vietnam. So, we must not only care deeply for what happens within ourselves, but also around us. From the view of a nation's people, it can be said, in another way, that we must be aware of what happens within our country and in our neighboring counterparts. Nowadays, the People Government of China has built many large dams alongside many hydroelectric facilities on the great Mekong River, destroying the habitats and causing negative effects in nations downstream.

In our presentations about the Mekong River (the Nine Dragon River in Vietnamese) at the UN's Vesak Summit from the 27th to the 30th of May, 2015 in Bangkok, Thailand, we indicated the origin of the Mekong from the Tibet Highland, running across six nations including China, Myanmar, Thailand, Laos, Cambodia, and Vietnam, stretching 4500 km long, making it the 12th longest river in the world. Within the past 20 years, there has been a hydroelectric exploitation program on the Mekong (Richard Cronin, 2010; Scott Pearse-Smith, 2012). By 2014, there were 26 hydroelectric dams on the main river, and 14 more on the Lancang River (the name of the Mekong's upstream in Yunnan, China). The barricades of the natural flow all have tremendous negative impacts on the natural habitats and have affected all six nations, especially the downstream ones such as Cambodia and Vietnam. We claimed that the hydroelectric dams were causing economic consequences, affecting the livelihood of millions of people living in the downstream nations. The flood of Mekong happens annually from June to October, with hundreds of deaths. The majority of the flood victims have been drowning children who lacked the supervision of their families' adults. In the Bangkok Declaration of the UN's Vesak Summit, the Mekong dilemma was referred to and a request was made to the countries in the ASEAN community and neighboring nations to join hands to solve the emergent situation of the Mekong River, as well as the ecological system.

Furthermore, aside from the large hydroelectric dams that China built, the authorities also constructed nuclear power plants situated close to Vietnam, e.g. the Fangchengang near the Quanzhou city in Guangxi, only 45 km (30 miles) away from Vietnam's border, and another nuclear power plant, Changjiang, West of Hainan Island. Although the nuclear power plants have not caused the current state of global warming, they do cause serious concerns regarding safety, security, and vulnerability to the nation of Vietnam. For if an incident was to happen in these Chinese plants near the Northern Vietnamese borders, there would not only be a disaster for the small common folks of China, but also the innocent people of Vietnam would be affected heavily as well. We carried out researches and concluded in the 'Effects of Chinese Nuclear Power Plants to the Safety of Homeland', that:

"(...) in summary, for the welfare of many people and many generations, we must be clearly aware of the danger[s] of the nuclear power source. And we must learn ways to prepare, prevent and provide for our families, communities and country, should unfortunate events manifest. The aforementioned Chinese nuclear plants, in particular, must be kept with a regular watchful eye to be ready to tackle and reduce the lethal as well as economic damages that would be dealt with by our people."

In this writing, our viewpoint was made clear that Vietnam was not ready to build and monitor nuclear power plants. We recommended, instead of nuclear or hydroelectric plants, that:

"A viable and stable alternative to power dams would be the sun and wind. Renewable technologies were proposed as possible strategies to develop the economy. The nations and peoples living

along the Mekong River are mostly Buddhists, the monasteries can play an important role in influencing the government's policies, and educate the people about the environmental costs and effects of these dams and the benefits of clean and renewable energy. These efforts can lower the conflicts in the future, economic and environment disasters and this great Buddhist River would not face a horrifying death".

Therefore, we must be aware that if the environmental dilemmas and global warming become worse, we and our children will face with unimaginable consequences.

To a child of Buddhism, Engaged Buddhism is especially a starting point. Engaged Buddhism was founded back in the 1960s in the Vietnam War by The Most Venerable Thich Nhat Hanh to kick-start the application of Buddhist knowledge into specific actions, bringing meditation and Buddhism into the common life to ease agonizing situations; from social to political, and from economic to environmental, for human beings and society.

According to two researchers, Queen, Chris and King, Sallie (1996), Engaged Buddhism has grown and developed and become quite popular in the West. We should bring the message of knowledge and love (empathy and wisdom) to relieve life from suffering. Mahayana entered life through positive efforts, (self-development, communal development; self-enlightenment, communal enlightenment). Buddhism chose the path of Middle Way. Poverty and illiteracy mean unhappiness, and unfulfilled basic needs mean difficulties in development. We need a true "eradication of hunger, reduction of poverty" and to improve the people's literacy as well as civil freedom, democracy, humanity, etc., so that our society can thrive, and we must have the awareness to protect and preserve our habitat in order to slow global warming, to conserve Mother Earth.

Generally, every human has basic needs. Everywhere, people want or try to have a joyful and happy life. We want a life of harmony, good health, and contentment - where we are not too worried, and can always care for future generations. Everyone knows that life is a temporary plane, and we only have one Mother Earth, home to 7 billion people living together. What remains is what we ought to do to savor this situation. So, we can do specific things in the time being in order to improve our life and the lives of the others.

The Solutions that can change the Carbon footprint:

I. For individuals and families:

1. Live a life of wisdom and moderation, and avoid excessive desires (frugal living).
2. Pick at least one day per week to practice vegetarianism/veganism.
3. Eat less meat or none at all (Meatless Monday).
4. Reduce, Reuse, and Recycle.
5. Bring yourself to make your living, working, and entertainment environment green, clean, and beautiful.

II. For the community/state:

1. Create policies/rules/programs to improve the eco-system.
2. Use eco-friendly or renewable energy.
3. Support healthy movements such as Living Green, Earth Day, etc.

4. Encourage and participate in healthy and socially-helpful organizations.
5. Grow gardens of vegetables, fruits for the community, participate in Farm-to-Fork programs, etc.

III. For the Commonwealth/Nation

1. Accept International conventions made to reduce the carbon footprint and to use renewable energy.
2. Developed countries should remove or reduce nuclear plants, avoid nuclear/chemical/biological warfare, etc.
3. Balance the distribution of food and wealth.
4. Eradicate dictatorial domination, and understand the mutual and correlative relationships in the universe.
5. For developing/undeveloped countries – for now, there should be no existence of nuclear plants because of the lack of morality, ability, and experiences.

Finally, this is the responsibility and duty that belongs to all of us in order to protect the only planet that humans are inhabiting. Moreover, every individual citizen of this world, especially the leaders, educators, politicians, etc., must consider it an ethical and moral responsibility to preserve Mother Earth. Therefore, we ought to start a new journey, striving to change life for ourselves, and for all those around us, into becoming better; and to leave a legacy of a clean, prosperous Earth for our descendants to inherit, including our own children.

Dr. Phe Bach, *Mira Loma High, SJUSD, Sacramento, CA.*
Dr. Khanh T. Tran, *AMI Environmental, USA.*

References:

1. Cox, P. M., Betts, R. A., Jones, C. D., Spall, S. A., & Totterdell, I. J. (2000). Acceleration of global warming due to carbon-cycle feedbacks in a coupled climate model. Nature, 408(6809), 184-187.

2. Hansen, J., Sato, M., Ruedy, R., Lacis, A., & Oinas, V. (2000). Global warming in the twenty-first century: An alternative scenario. Proceedings of the National Academy of Sciences, 97(18), 9875-9880.

3. Schuldt, J. P., Konrath, S. H., & Schwarz, N. (2011). "Global warming" or "climate change"? Whether the planet is warming depends on question wording. Public Opinion Quarterly, nfq073. http://worldpopulationreview.com/countries/vietnam-population/

4. Lorenzoni, I., & Pidgeon, N. F. (2006). Public views on climate change: European and USA perspectives. Climatic change, 77(1-2), 73-95.

5. Queen, C. S., & King, S. B. (1996). Engaged Buddhism: Buddhist Liberation Movements in Asia. New York: Albany State University Press. p. 2. ISBN 0-7914-2843-5.

6. Thích Nguyên Hiệp, Đạo Đức Học Phật Giáo Và Vấn Đề Môi Trường. Thư Viện Hoa Sen. Tải xuống ngày 27 tháng 10, 2015. http://thuvienhoasen.org/a4365/dao-duc-hoc-phat-giao-va-van-de-moi-truong-thich-nguyen-hiep

7. Thích Tâm Pháp, Phật Giáo và Môi Trường trong Tuyển tập Phật Thành Đạo. Nhiều tác giả, ở Phần III. Tải xuống ngày 20 tháng 10, 2015. http://www.tuvienquangduc.com.au/DucPhat/40td-tamphap.html

8. Time for Change. Cause and effect for global warming. Tải xuống ngày 10 tháng 10, 2015. http://timeforchange.org/cause-and-effect-for-global-warming

9. Trần Tiễn Khanh và Bạch X. Phẻ (2015), Ảnh hưởng nhà máy điện hạt nhân của trung quốc và sự an nguy của tổ quốc. Phe Bach's Blog. Tải xuống ngày 20 tháng 10, 2015. http://phebach.blogspot.com/2015/06/anh-huong-nha-may-ien-hat-nhan-cua.html

10. Walther, G. R., Post, E., Convey, P., Menzel, A., Parmesan, C., Beebee, T. J., ... & Bairlein, F. (2002). Ecological responses to recent climate change. Nature, 416(6879), 389-395.

11. World Population Review (2015). Vietnam Population 2015.

ZEN MASTER THICH NHAT HANH: ONLY LOVE CAN SAVE US FROM CLIMATE CHANGE

Jo Confino/The Guardian
Mon 21 Jan 2013 07.31 EST

Leading spiritual teacher warns that if people cannot save themselves from their own suffering, how can they be expected to worry about the plight of Mother Earth.

Zen master Thich Nhat Hanh, one of the world's leading spiritual teachers, is a man at great peace even as he predicts the possible collapse of civilisation within 100 years as a result of runaway climate change.

The 86-year-old Vietnamese monk, who has hundreds of thousands of followers around the world, believes the reason most people are not responding to the threat of global warming, despite overwhelming scientific evidence, is that they are unable to save themselves from their own personal suffering, never mind worry about the plight of Mother Earth.

Thay, as he is known, says it is possible to be at peace if you pierce through our false reality, which is based on the idea of life and death, to touch the ultimate dimension in Buddhist thinking, in which energy cannot be created or destroyed.

By recognising the inter-connectedness of all life, we can move beyond the idea that we are separate selves and expand our compassion and love in such a way that we take action to protect the Earth.

Look beyond fear

In Thay's new book, Fear, he writes about how people spend much of their lives worrying about getting ill, ageing and losing the things they treasure most, despite the obvious fact that one day they will have to let them all go.

When we understand that we are more than our physical bodies, that we didn't come from nothingness and will not disappear into nothingness, we are liberated from fear, he says; fearlessness is not only possible but the ultimate joy.

"Our perception of time may help," Thay told me in his modest home in Plum Village monastery near Bordeaux. "For us it is very alarming and urgent, but for Mother Earth, if she suffers she knows she has the power to heal herself even if it takes 100m years. We think our time on earth is only 100 years, which is why we are impatient. The collective karma and ignorance of our race, the collective anger and violence will lead to our destruction and we have to learn to accept that.

"And maybe Mother Earth will produce a great being sometime

in the next decade ... We don't know and we cannot predict. Mother Earth is very talented. She has produced Buddhas, bodhisattvas, great beings.

"So take refuge in Mother Earth and surrender to her and ask her to heal us, to help us. And we have to accept that the worst can happen; that most of us will die as a species and many other species will die also and Mother Earth will be capable after maybe a few million years to bring us out again and this time wiser."

Confront the truth

Thay suggests that our search for fame, wealth, power and sexual gratification provides the perfect refuge for people to hide from the truth about the many challenges facing the world. Worse still, our addiction to material goods and a hectic lifestyle provides only a temporary plaster for gaping emotional and spiritual wounds, which only drives greater loneliness and unhappiness.

Thay, who has just celebrated the 70th anniversary of his ordination, reflects on the lack of action over the destruction of ecosystems and the rapid rate of biodiversity loss: "When they see the truth it is too late to act ... but they don't want to wake up because it may make them suffer. They cannot confront the truth. It is not that they don't know what is going to happen. They just don't want to think about it.

"They want to get busy in order to forget. We should not talk in terms of what they should do, what they should not do, for the sake of the future. We should talk to them in such a way that touches their hearts, that helps them to engage on the path that will bring them true happiness; the path of love and understanding, the courage to let go. When they have tasted a little bit of peace and love, they may wake up."

Thay created the Engaged Buddhism movement, which promotes the individual's active role in creating change, and his mindfulness training – an ethical roadmap – calls on practitioners to boycott products that damage the environment and to confront social injustice.

Given the difficulty of convincing those with vested interests to change their behaviour, Thay says a grassroots movement is necessary, citing the tactics used by Gandhi, but insists that this can be effective only if activists first deal with their own anger and fears, rather than projecting them onto those they see at fault.

Awakened consumers can influence how companies act

On companies that produce harmful products, he says: "They should not continue to produce these things. We don't need them. We need other kinds of products that help us to be healthier. If there is awakening in the ranks of consumers, then the producer will have to change. We can force him to change by not buying.

"Gandhi was capable of urging his people to boycott a number of things. He knew how to take care of himself during non-violent operations. He knew how to preserve energy because the struggle is long, so spiritual practice is very much needed in an attempt to help change society."

Thay, the author of more than 100 books, including the best-selling Miracle of Mindfulness, says that while it is difficult for those holding the strings of power to speak out against the destructive nature of the current economic system, for fear of being ostracised and ridiculed, we do need more leaders to have the courage to challenge the status quo.

For business and political leaders to do that, they need to cultivate compassion in order to embrace and diminish the ego, Thay says.

"You have the courage to do it [speak out] because you have compassion, because compassion is a powerful energy," he says. "With compassion you can die for other people, like the mother who can die for her child. You have the courage to say it because you are not afraid of losing anything, because you know that understanding and love is the foundation of happiness. But if you have fear of losing your status, your position, you will not have the courage to do it."

A moment of contemplation

While many people are becoming disorientated by the complexity of their lives and by the overwhelming array of choices offered by our consumer society, Thay's retreats offer a profoundly simple alternative.

Over Plum Village's three-month winter retreat, Thay repeatedly instructs the hundreds of monks, nuns and lay practitioners about switching off the non-stop noise in their heads and focusing on the core of mindfulness; the joy of breathing, of walking, of contemplation in the present moment.

Rather than searching for answers to life in the study of philosophy, or seeking adrenaline charged peak experiences, Thay suggests that true happiness can be found by touching the sacred in the very ordinary experiences of life, which we largely overlook.

How often do we fully appreciate, for example, how hard our hearts work day and night to keep us alive? He suggests it is possible to discover profound truths through concentrating on something as basic as eating a carrot, as you get the insight that the vegetable cannot exist without the support of the entire universe.

"If you truly get in touch with a piece of carrot, you get in touch with the soil, the rain, the sunshine," he says. "You get in touch with Mother Earth and eating in such a way, you feel in touch with true life, your roots, and that is meditation. If we chew every morsel of our food in that way we become grateful and when you are grateful, you are happy."

Despite meditating every day for the past seven decades, Thay believes there is still much to learn. "In Buddhism we speak of love as something limitless," he says. "The four elements of love which are loving kindness, compassion, joy and equanimity, have no frontiers.

"Buddha is thinking like that. His followers call him the perfect one but that is out of love, for the truth is you can never be perfect. But we don't need to be perfect. That is a good thing to know. If you make a little bit of progress every day, a little bit more joy and peace, that is good enough so Thay continues to practice and his insight grows deeper every day.

"There is no limit of the practice. And I think that is true of the human race. We can continue to learn generation after generation and now is time to begin to learn how to love in a non-discriminatory way because we are intelligent enough, but we are not loving enough as a species."

Thich Nhat Hanh: a life lived away from the public eye

Thay is often compared to the Dalai Lama but has largely escaped the public's gaze, deciding to live the life of a simple monk. He has avoided the trap of being surrounded by celebrities and will give interviews only to journalists who have spent time beforehand meditating with him on the basis that mindfulness needs to be experienced, rather than described.

But Thay is no wallflower and has led an extraordinary life, including a nomination for the Nobel peace prize from Martin Luther King in 1967 for his work in seeking an end to the Vietnam war. In his nomination King said: "I do not personally know of anyone more worthy of [this prize] than this gentle monk from Vietnam. His ideas for peace, if applied, would build a monument to ecumenism, to world brotherhood, to humanity".

Thay set up Plum Village 30 years ago after being exiled from his home country and has since added monasteries in Thailand, Hong Kong and the US, as well as an applied Buddhist institute in Germany. He has continued to work for peaceful solutions to conflicts around the world, including holding several retreats for Israelis and Palestinians.

In 2009 he faced conflict in his own life, when the Vietnamese authorities closed down his recently opened monastery at Bhat Nha after a campaign of harassment and violence. Thay believes the action, which sparked an outcry from the EU and other countries, was orchestrated by the Chinese following his public support for Tibet. The 400 monks and nuns were dispersed but still operate quietly within the country.

The Guardian's release of US embassy cables highlighted concern about the crackdown. One confidential cable said: "Vietnam's poor handling of the situations at the Plum Village community at the Bat Nha Pagoda and the Dong Chiem Catholic parish last week – particularly the excessive use of violence – is troublesome and indicative of a larger GVN crackdown on human rights in the run-up to the January 2011 Party Congress"

Despite all his achievements, including a recent stint as guest editor at the Times of India, Thay is modest when he looks back at his life.

"There is not much we have achieved except some peace, some contentment inside. It is already a lot," he says. "The happiest moments are when we sit down and we feel the presence of our brothers and sisters, lay and monastic, who are practicising walking and sitting mediation. That is the main achievement and other things like publishing books and setting up institutions like in Germany, they are not important.

"It is important we have a sangha [community] and the insight came that the Buddha of our time may not be an individual but it might be a sangha. If every day you practice walking and sitting meditation and generate the energy of mindfulness and concentration and peace, you are a cell in the body of the new Buddha. This is not a dream but is possible today and tomorrow. The Buddha is not something far away but in the here and in the now."

While Thay is still in good health and sharp as a pin, he is not getting any younger and may soon begin to start pulling back from the strenuous schedule that has seen him repeatedly criss-crossing the world, leading retreats and passing on his teachings. This year he travels across the US and Asia – perhaps his last major foreign trip.

Given his belief in no birth and no death, how does he feel about his own passing?

"It is very clear that Thay will not die but will continue in other people," he says. "So there is nothing lost and we are happy because we are able to help the Buddha to renew his teaching. He is deeply misunderstood by many people so we try to make the teaching available and simple enough so that all people can make good use of that teaching and practice."

As he lifts a glass of tea to drink, he adds: "I have died already many times and you die every moment and you are reborn in every moment so that is the way we train ourselves. It is like the tea. When you pour the hot water in the tea, you drink it for the first time, and then you pour again some hot water and you drink, and after that the tea leaves are there in the pot but the flavour has gone into the tea and if you say they die it is not correct because they continue to live on in the tea, so this body is just a residue.

"It still can provide some tea flavour but one day there will be no tea flavour left and that is not death. And even the tea leaves, you can put them in the flower pot and they continue to serve so we have to look at birth and death like that. So when I see young monastics and lay people practicing, I see that is the continuation of the Buddha, my continuation."

Prompted by a letter that informed him that someone has built a temple in Hanoi to commemorate his life, Thay recently sent a letter to the Tu Hieu temple in central Vietnam, where he trained as a novice monk, making it clear he does not want a shrine built in his honour when he dies: "I said don't waste the land of the temple in order to build me a stupha. Do not put me in a small pot and put me in there. I don't want to continue like that. It is better to put the ash outside to help the trees to grow. That is a meditation."

He adds: "I recommend that they make the inscription outside on the front 'I am not in here'. And then if people do not understand, you add a second sentence 'I am not out there either' and if still they don't understand on the third and the last; 'I may be found maybe in your way of breathing or walking.'"

Jo Confino/The Guardian
Mon 21 Jan 2013 07.31 EST

THIS BUDDHIST MONK
IS AN UNSUNG HERO
IN THE WORLD'S CLIMATE
FIGHT

Jo Confino/Huffpost
01/22/2016 04:04 pm ET

The architect of the historic Paris climate negotiations credits the teachings of Thich Nhat Hanh with helping broker the deal.

One of the guiding forces behind the scenes of the Paris climate agreement is an 89-year-old Vietnamese Zen Buddhist monk.

Christiana Figueres, who led the climate talks, has credited Thich Nhat Hanh with having played a pivotal role in helping her to develop the strength, wisdom and compassion needed to forge the unprecedented deal backed by 196 countries.

Figueres, the executive secretary of the United Nations Framework Convention on Climate Change, says the teachings of Thay, as he is known to his hundreds of thousands of followers around the world, "literally fell into my lap" when she was going through a deep personal crisis three years ago.

She says the Buddhist philosophy of Thay, who is currently recovering from a serious stroke, helped her to deal with the crisis while also allowing her to maintain her focus on the climate talks.

Figueres said she realized that "I have to have something here, because otherwise I can't deal with this and do my job, and it was very clear to me that there was no way that I could take a single day off," she told The Huffington Post this week at the World Economic Forum's annual meeting in Davos, Switzerland.

"This has been a six-year marathon with no rest in between," she said. "I just really needed something to buttress me, and I don't think that I would have had the inner stamina, the depth of optimism, the depth of commitment, the depth of the inspiration if I had not been accompanied by the teachings of Thich Nhat Hanh."

So what did Thay teach her?

Figueres illustrates this via a visit she made to his monastery in Waldbrol, Germany, which was once a mental institution with 700 patients, before the Nazis came along to exterminate them and took over the premises for the Hitler Youth.

She says Thay chose to locate his monastery there "because he wanted to prove that it is completely possible to turn pain into love, to turn being a victim into being a victor, to turn hate into love and forgiveness, and he was intent in showing that in this place that had been associated with such absolute, inhuman cruelty."

"The first thing that he did was he wrote to the Buddhist community and he said, 'I want hearts. I want hand-sewn hearts, one for each of the patients who were killed here, so that we can begin to transform this building, and this space, and this energy,'" Figueres told HuffPost.

"It was such a powerful story for me, right? Because in many ways, that is the journey that we have been on in the climate negotiations," she continued. "It is a journey from blaming each other, to actually collaborating. It's a journey from feeling completely paralyzed, helpless, exposed to the elements, to actually feeling empowered that we can do this."

"It's actually been for me internally a beautiful journey of healing. So for me, I've sort of been living life at many different levels, because I had to turn my own personal crisis, I had to transform that," Figueres went on to say. "I'm still in the midst of that, I'm not going to say I'm way over on the other side, but I had to do that for myself."

"I felt this is exactly the energy that the climate change convention negotiations need, all inspired, you know, by this amazing teaching," she said.

In fact when Thay arrived for the first time at the former Nazi headquarters, which has 400 rooms, he wrote a letter to the patients who died, which is read every day at the monastery by the monks and nuns who live there.

"Now the Sangha [community] has come, the Sangha has heard and understood your suffering and the injustice you endured," it says. "The people who caused your suffering have also suffered a lot. They did not know what they were doing at that time. So please allow compassion and forgiveness to be born in your heart so that they also can have a chance to transform and heal. Please support the Sangha and the next many generations of practitioners so that we can transform these places of suffering into places of transformation and healing, not only for Waldbrol but for the whole country of Germany and the world."

Thay, who is considered by many to be the father of mindfulness in the West and has been an environmentalist activist for more than two decades, has other monasteries around the world and has built the fastest-growing monastic order in the world. He is also well respected by senior leaders across the United States.

I don't think I would have had the inner stamina, the depth of optimism, the depth of commitment, the depth of the inspiration if I had not been accompanied by the teachings of Thich Nhat Hanh."U.N. climate chief Christiana Figueres

Last year, he was invited by the World Bank president, Jim Yong Kim, to the organization's Washington headquarters for an event with the staff. Kim's favorite book is Thay's *The Miracle of Mindfulness,* and he praises the Zen monk's practice for being "deeply passionate and compassionate toward those who are suffering."

Thay visited Silicon Valley in 2013 at the invitation of Google and was also asked to lead a private day of mindfulness for CEOs of 15 of the world's most powerful technology companies.

Marc Benioff, CEO of cloud computing giant Salesforce, has been actively supporting Thay's rehabilitation after he fell ill.

Thay has led an extraordinary life, including a nomination for the Nobel peace prize from Martin Luther King in 1967 for his work in seeking an end to the Vietnam war. In his nomination King said: "I do not personally know of anyone more worthy of [this prize] than this gentle monk from Vietnam. His ideas for peace, if applied, would build a monument to ecumenism, to world brotherhood, to humanity."

Jo Confino/Huffpost
01/22/2016 04:04 pm ET

THICH NHAT HANH'S STATEMENT ON CLIMATE CHANGE FOR THE UNITED NATIONS

Thich Nhat Hanh/Plum Village
2nd July, 2015

In 2014, the United Nations Framework Convention on Climate Change (UNFCCC) approached Thich Nhat Hanh as an international faith leader, to request a brief statement about climate change and our relationship to one another and to the Earth. This statement was published on the UNFCCC website, ahead of the Paris Climate Summit in September 2015.

Falling in Love with the Earth

This beautiful, bounteous, life-giving planet we call Earth has given birth to each one of us, and each one of us carries the Earth within every cell of our body.

We and the Earth are one

The Earth is our mother, nourishing and protecting us in every moment–giving us air to breathe, fresh water to drink, food to eat and healing herbs to cure us when we are sick. Every breath we inhale contains our planet's nitrogen, oxygen, water vapor and trace elements. When we breathe with mindfulness, we can experience our interbeing with the Earth's delicate atmosphere, with all the plants, and even with the sun, whose light makes possible the miracle of photosynthesis. With every breath we can experience communion. With every breath we can savor the wonders of life.

We need to change our way of thinking and seeing things. We need to realise that the Earth is not just our environment. The Earth is not something outside of us. Breathing with mindfulness and contemplating your body, you realise that you are the Earth. You realise that your consciousness is also the consciousness of the Earth. Look around you–what you see is not your environment, it is you.

Great Mother Earth

Whatever nationality or culture we belong to, whatever religion we follow, whether we're Buddhists, Christians, Muslims, Jews, or atheists, we can all see that the Earth is not inert matter. She is a great being, who has herself given birth to many other great beings–including buddhas and bodhisattvas, prophets and saints, sons and daughters of God and humankind. The Earth is a loving mother, nurturing and protecting all peoples and all species without discrimination.

When you realize the Earth is so much more than simply your environment, you'll be moved to protect her in the same way as you would yourself. This is the kind of awareness, the kind of awakening that we need, and the future of the planet depends on whether we're able to cultivate this insight or not. The Earth and all species on Earth are in real danger. Yet if we can develop a deep relationship with the Earth, we'll have enough love, strength and awakening in order to change our way of life.

Falling in love

We can all experience a feeling of deep admiration and love when we see the great harmony, elegance and beauty of the Earth. A simple branch of cherry blossom, the shell of a snail or the wing of a bat – all bear witness to the Earth's masterful creativity. Every advance in our scientific understanding deepens our admiration and love for this wondrous planet. When we can truly see and understand the Earth, love is born in our hearts. We feel connected. That is the meaning of love: to be at one.

Only when we've truly fallen back in love with the Earth will our actions spring from reverence and the insight of our interconnectedness. Yet many of us have become alienated from the Earth. We are lost, isolated and lonely. We work too hard, our lives are too busy, and we are restless and distracted, losing ourselves in consumption. But the Earth is always there for us, offering us everything we need for our nourishment and healing: the miraculous grain of corn, the refreshing stream, the fragrant forest, the majestic snow-capped mountain peak, and the joyful birdsong at dawn.

True Happiness is made of love

Many of us think we need more money, more power or more status before we can be happy. We're so busy spending our lives chasing after money, power and status that we ignore all the conditions for happiness already available. At the same time, we lose ourselves in buying and consuming things we don't need, putting a heavy strain on both our bodies and the planet. Yet much of what we drink, eat, watch, read or listen to, is toxic, polluting our bodies and minds with violence, anger, fear and despair.

As well as the carbon dioxide pollution of our physical environment, we can speak of the spiritual pollution of our human environment: the toxic and destructive atmosphere we're creating with our way of consuming. We need to consume in such a way that truly sustains our peace and happiness. Only when we're sustainable as humans will our civilization become sustainable. It is possible to be happy in the here and the now.

We don't need to consume a lot to be happy; in fact we can live very simply. With mindfulness, any moment can become a happy moment. Savoring one simple breath, taking a moment to stop and contemplate the bright blue sky, or to fully enjoy the presence of a loved one, can be more than enough to make us happy. Each one of us needs to come back to reconnect with ourselves, with our loved ones and with the Earth. It's not money, power or consuming that can make us happy, but having love and understanding in our heart.

The bread in your hand is the body of the cosmos

We need to consume in such a way that keeps our compassion alive. And yet many of us consume in a way that is very violent. Forests are cut down to raise cattle for beef, or to grow grain for liquor, while millions in the world are dying of starvation. Reducing the amount of meat we eat and alcohol we consume by 50% is a true act of love for ourselves, for the Earth and for one another. Eating with compassion can already help transform the situation our planet is facing, and restore balance to ourselves and the Earth.

Nothing is more important than brotherhood and sisterhood

There's a revolution that needs to happen and it starts from inside each one of us. We need to wake up and fall in love with Earth. We've been homo sapiens for a long time. Now it's time to become homo conscius. Our love and admiration for the Earth has the power to unite us and remove all boundaries, separation and discrimination. Centuries of individualism and competition have brought about tremendous destruction and alienation. We need to re-establish true communication–true communion–with ourselves, with the Earth, and with one another as children of the same mother. We need more than new technology to protect the planet. We need real community and co-operation.

All civilisations are impermanent and must come to an end one day. But if we continue on our current course, there's no doubt that our civilisation will be destroyed sooner than we think. The Earth may need millions of years to heal, to retrieve her balance and restore her beauty. She will be able to recover, but we humans and many other species will disappear, until the Earth can generate conditions to bring us forth again in new forms. Once we can accept the impermanence of our civilization with peace, we will be liberated from our fear. Only then will we have the strength, awakening and love we need to bring us together. Cherishing our precious Earth–falling in love with the Earth–is not an obligation. It is a matter of personal and collective happiness and survival.

Thich Nhat Hanh/Plum Village

BUDDHIST CLIMATE CHANGE STATEMENT TO WORLD LEADERS 2015

GLOBAL BUDDHIST CLIMATE CHANGE COLLECTIVE (GBCCC)

October 29th, 2015
(Signatures amended: November 29th, 2015)

We, the undersigned Buddhist leaders, come together prior to the 21st Session of the Conference of Parties (COP21) to the UN Framework Convention on Climate Change (UNFCCC) in Paris, in order to add our voices to the growing calls for world leaders to cooperate with compassion and wisdom and reach an ambitious and effective climate agreement.

We are at a crucial crossroads where our survival and that of other species is at stake as a result of our actions. There is still time to slow the pace of climate change and limit its impacts, but to do so, the Paris summit will need to put us on a path to phase out fossil fuels. We must ensure the protection of the most vulnerable, through visionary and comprehensive mitigation and adaptation measures.

Our concern is founded on the Buddha's realization of

dependent co-arising, which interconnects all things in the universe. Understanding this interconnected causality and the consequences of our actions are critical steps in reducing our environmental impact. Cultivating the insight of interbeing and compassion, we will be able to act out of love, not fear, to protect our planet. Buddhist leaders have been speaking about this for decades. However, everyday life can easily lead us to forget that our lives are inextricably interwoven with the natural world through every breath we take, the water we drink, and the food we eat. Through our lack of insight, we are destroying the very life support systems that we and all other living beings depend on for survival.

We believe it imperative that the global Buddhist community recognize both our dependence on one another as well as on the natural world. Together, humanity must act on the root causes of this environmental crisis, which is driven by our use of fossil fuels, unsustainable consumption patterns, lack of awareness, and lack of concern about the consequences of our actions.

We strongly support "The Time to Act is Now: A Buddhist Declaration on Climate Change," which is endorsed by a diverse and global representation of Buddhist leaders and Buddhist sanghas. We also welcome and support the climate change statements of other religious traditions. These include Pope Francis's encyclical earlier this year, *Laudato Si': On Care for Our Common Home*, the *Islamic Declaration on Climate Change*, as well as the upcoming *Hindu Declaration on Climate Change*. We are united by our concern to phase out fossil fuels, to reduce our consumption patterns, and the ethical imperative to act against both the causes and the impacts of climate change, especially on the world's poorest.

To this end, we urge world leaders to generate the political will to close the emissions gap left by country climate pledges and ensure that the global temperature increase remains below 1.5 degrees Celsius, relative to pre-industrial levels. We also ask for a common commitment to scale up climate finance, so as to help developing countries prepare for climate impacts and to help us all transition to a safe, low carbon future.

The good news is that there is a unique opportunity at the Paris climate negotiations to create a turning point. Scientists assure us that limiting the rise in the global average temperature to less than 1.5 degrees Celsius is technologically and economically feasible. Phasing out fossil fuels and moving toward 100 percent renewable and clean energy will not only spur a global, low-carbon transformation, it will also help us to embark on a much-needed path of spiritual renewal. In addition to our spiritual progression, in line with UN recommendations, some of the most effective actions individuals can take are to protect our forests, move toward a plant-based diet, reduce consumption, recycle, switch to renewables, fly less, and take public transport. We can all make a difference.

We call on world leaders to recognize and address our universal responsibility to protect the web of life for the benefit of all, now and for the future.

For these reasons, we call on all Parties in Paris:

To be guided by the moral dimensions of climate change as indicated in Article 3 of the United Nations Framework Convention on Climate Change (UNFCCC).

To agree to phase out fossil fuels and move towards 100 percent renewables and clean energy.

To create the political will to close the emissions gap left by country climate pledges so as to ensure that the global temperature increase remains below 1.5 degrees Celsius, relative to pre-industrial levels.

To make a common commitment to increase finance above the US$100 billion agreed in Copenhagen in 2009, including through the Green Climate Fund (GCF), to help vulnerable developing countries prepare for climate impacts and transition towards a low-carbon economy.

The time to act is now.

Yours sincerely,

Global Buddhist Leaders

His Holiness the Dalai Lama Tenzin Gyatso,
14th Dalai Lama

Zen Master Thích Nhất Hạnh,
Patriarch of the Plum Village International Community
of Engaged Buddhists

His Holiness the 17th Gyalwang Karmapa,
Head of the Karma Kagyu Tradition

Eminent Buddhist Monastic Leaders.
Alphabetically By Country

His Holiness Dr. Dharmasena Mahathero,

The Supreme Patriarch (Sangharaja)
of the Bangladesh Sangha, Bangladesh

Ven. Lama Lobzang,
Secretary General
of the International Buddhist Confederation (IBC), India

Most Ven. Jetsunma Tenzin Palmo,
President of Sakyadhita International Association
of Buddhist Women, India

His Eminence Jaseung Sunim,
President, Jogye Order of Korean Buddhism, Korea

Most Ven. Myeong Seong Sunim,
President Emeritus
of the Korean National Bhikkhuni Association, Korea

Ven. B. Sri Saranankara Nayaka Maha Thera,
Chief Adhikarana Sangha Nayaka
of Malaysia, Kuala Lumpur, Malaysia

His Eminence Rev. Khamba Lama Gabju Demberel,
The Supreme Head of Mongolian Buddhists, Mongolia

His Holiness Dr. Bhaddanta Kumarabhivamsa,
Sangharaja,
and Chairman State Sangha Maha Nāyaka Committee, Myanmar

**His Eminence Agga Maha Panditha Dawuldena Gnanissara
Maha Nayaka Thera**,
Mahanayaka Thero, The Supreme Prelate
of the Amarapura Maha Nikaya, Sri Lanka

Dharma Master Cheng Yen,

Buddhist Tzu Chi Foundation, Taiwan

Most Ven. Aryawangso,
Vice President of the World Fellowship of Buddhists, Thailand

His Holiness Thích Phổ Tuệ, Supreme Patriarch of All Vietnam
Buddhist Sangha, Vietnam

Most Ven. Thích Nữ Tịnh Nguyện,
President of Vietnam Bhikkhuni Sangha, Vietnam

Ven. Bhikkhu Bodhi,
President, Buddhist Association of the USA

Ven. Tathaloka Theri,
Founder, North American Bhikkhuni Association, USA

Eminent Buddhist Leaders,
Alphabetically By Country

Her Royal Highness Ashi Kesang Wangmo Wangchuk,
Bhutan

Rev. Olivier Reigen Wang-gen,
President, Buddhist Union of France (UBF), France

Rev. Hakuga Murayama,
18th President and International Chairperson,
All Japan Young Buddhist Association (JYBA), Japan

Yoshitaka Oba,
General Director, Soka Gakkai International (SGI), Japan

Rev. Sensei Amala Wrightson,
Buddhist Climate Change Leader, New Zealand

Dr. A.T. Ariyaratne,

Founder of Sarvodaya Shramadana, Sri Lanka

Jamie Creswell,
President of the European Buddhist Union, UK

Rev. Roshi Dr. Joan Jiko Halifax,
Deputy Secretary General
of the International Buddhist Confederation, USA

www.ingramcontent.com/pod-product-compliance
Lightning Source LLC
Chambersburg PA
CBHW031341060726
47590CB00007B/2581